For their inspiration and exactitude: Chris Quigg, Marc Gillinov, Andrei Severny, Jeana Monteiro, Ken Carbone, Michael Racz, Seth Powsner, Howard Gralla, Wendy MacNaughton, Maira Kalman, Ken Kocienda, Duke Cameron, David S. Smith, Daryl Morey, Earl Devaney, Dan Snow, Jackie Heinricher, Kevin Roche, Austin Kleon, Michael Fogleman, Adam Schwartz, Colleen Bushell, Mike Bostock, Jonathan Corum, Alyssa Goodman, Maya Lin, Randall Monroe (xkcd), Bret Victor, Stewart Brand, Maria Popova, David Donoho, Stephen Stigler, Sister Helen Prejean, Max Roser, Inge Druckrey, Paul Ekman, Philip Greenspun, Nicholas Cox, Graham Larkin, and my twitter friends Elisabeth Bik, Darrel Francis, Frank Harrell, John Mandrola, Vinay Prasad, Maarten van Smeden, Eric Topol, Gershon Klapper.

For making this book happen, my colleagues Kristie Addy, Cynthia Bill, Andy Conklin, Bonnie Darling, Dawn Finley, John Fournier, Mahmoud Hamadani, Brian Kelly, Emily Kirkegaard, Elaine Morse, Pam Mozier, Jared Ocoma, Peter Taylor, Michael Tracy, Carolyn Williams, Tom Woodward.

Parts of this book were presented in talks at the American Association of Thoracic Surgery, MIT Sloan Sports Analytics Conference, Microsoft, ValueAct Capital, U.N. Data Viz Camp, Fermi National Accelerator Laboratory, Morgan Stanley, University of Twente (Netherlands), Pixar, D. E. Shaw Research, Princeton University, National Center for Supercomputing Applications (University of Illinois), Facebook, ET Modern Gallery, and in my one-day course 'Presenting Data and Information.' Research, analysis, design, publication, were funded by my Graphics Press LLC, Hogpen Hill Farms, ET Modern Gallery.

Printed by GHP on Sappi Flo Text, binding by Superior Packaging and Finishing.

For my mother, Professor Virginia James Tufte, 1918–2020

Endlessly curious, watching an eclipse of the sun through solar observation glasses, and at Pebble Beach, California. Author of 6 books, including *Artful Sentences: Syntax as Style.*

SEEING WITH FRESH EYES
MEANING SPACE DATA TRUTH
EDWARD TUFTE

THE THINKING EYE

To see the ordinary so intensely

that the ordinary becomes extraordinary, becoming

so focused, so specific about something,

that it becomes something other than what it ordinarily is

Always on, thinking eyes see intensely actively skeptically,

scan globally focus locally

see at varying scales of space and time, approximating

ways through multiplicity, detecting

how things happen, move, act, interact, seeing

with fresh eyes vacation eyes unhindered

by self-confirming words, models, expectations

not seeing something different is not seeing anything at all

Grace Hopper saying The most dangerous phrase in the language is

'We've always done it this way'

Staying in optical experiences, forgetting the name of what one sees

laughing playful eyes shut up and look if you see something, say nothing

defamiliarize decontextualize recontextualize reform remodel

Thinking eyes are of this world empirical specific practical self-aware

asking disbelieving challenging making the familiar unfamiliar

you

How do I really know that?

they

you

How could I possibly ever know that?

they

Thinking eyes reason intensely about what they see

reason about verbs links mechanisms connections dynamics

reason about what things do not what things are named

reason across multiple time-horizons then now forever

name re-name re-model

Thinking eyes compare model choose doubt decide compare again

Thinking eyes act make something of seeing and reasoning

discover produce construct

write a report make an artwork teach a class have an insight

understand explain show get on with it

To produce, construct, model, remodel,

to act is essential, it is the difference between

spectator/player consumer/producer artchat vs. artwork

anecdote vs. evidence process/outcome retrospective/prospective

presentations pitching vs. demonstrations comparing

Craig Venter saying 'Good ideas are a dime a dozen for a smart person,

what distinguishes good from great is how an idea is executed –

how it becomes reality'

Thinking eyes identify/know/celebrate

excellence forever universal knowledge

gathering consequences

staying in place beyond memory's imprecision

seeing learning doing doubting

are the meaning of intelligent life

1 MEANING AND SPACE: SEE WITH FRESH EYES, QUESTION EVERYTHING, REMODEL CONVENTIONAL MODELS

WHAT DOES SPACE DO AND MEAN? AND WHERE?

How beautiful it was then, through that void, to draw lines and parabolas,
pick out the precise point, the intersection between space and time
when the event would spring forth, undeniable in the prominence of its glow.

ITALO CALVINO

music is the space between the notes
it's not the notes you play
it's the notes you don't play

MILES DAVIS

the horizon is not a line
it is a space

JOANNE CHEUNG

air is a material
seen worked transformed seen
just like steel stone earth

ARTWORKERS, WALLERS, ARCHITECTS

A vessel is useful only through its emptiness. It is the space opened in a wall
that serves as a window. Thus it is the nonexistent in things which makes them serviceable.

LAO TSE

Communications are living or dead depending upon organization of their blank spaces.
A single character gains clarity and meaning by orderly relationships of the surrounding space backgrounds.

GYÖRGY KEPES

This subject – turning the line in a poem – is one that every poet deals with throughout his or
her working life. Every turning is a meaningful decision, the effect of which is sure to be felt by the reader.
I cannot say too many times how powerful the techniques of line length and line breaks are.

MARY OLIVER

Vacuum, in modern physics, is what you get when you remove everything you can,
whether practically or in principle. Alternatively, vacuum is the state of minimum energy.
Intergalactic space is a good approximation to a vacuum.

Void, on the other hand, is a theoretical idealization. It means nothingness:
space without independent properties, whose only role is to keep everything
from happening in the same place. Void gives subatomic particles addresses, nothing more.

FRANK WILCZEK

You can't say A is made of B or vice versa. All mass is interaction.

RICHARD FEYNMAN

there is no there there

GERTRUDE STEIN

make sure your code 'does nothing' gracefully

THE ELEMENTS OF PROGRAMMING STYLE

nothing that is not there and the nothing that is

WALLACE STEVENS

'What are you rebelling against?'
'What do you got?'

MARLON BRANDO, THE WILD ONE

SPACE AND ITS COUNTERPARTS: A SPACE OF SPACES

space	meaning		meaning	space
emptiness	fullness		fullness	emptiness
absence	presence		presence	absence
void	material		material	void
silence	sound		sound	silence
ground	figure	reside together concrete and intrinsic	figure	ground
negative space	positive space		positive space	negative space
air space	material space		material space	air space
stillness	motion		motion	stillness
transparent	opaque		opaque	transparent
shadow	light		light	shadow

THINKING OFF THE GRID, THINKING SOMBREROS

Rationalists, wearing square hats,
Think, in square rooms,
Looking at the floor,
Looking at the ceiling.
They confine themselves
To right-angled triangles.
If they tried rhomboids,
Cones, waving lines, ellipses –
As, for example, the ellipse of the half-moon –
Rationalists would wear sombreros.

WALLACE STEVENS

SIX SIGNIFICANT LANDSCAPES, VI 1916

In mounting shows at the Guggenheim, artworks are positioned by eye – rather than by electronic laser-levels, squares, plumb-bobs – because viewers walking on the curved and sloped ramp are out of plumb, and the museum space is fully sombrero.

GUGGENHEIM MUSEUM GIACOMETTI EXHIBIT ET 2018

Models summarize, show, and explain something relevant. Their purpose is to cause consequential actions. Some models are better than others. Models in science and engineering are special, because they are based on Nature's forever universal laws, expressed in mathematics describing the physical world. Lacking the truth guarantee of Nature's mathematical laws, the human sciences are much harder than rocket science.

To choose a model is to choose assumptions – unknown, unseen, forgotten. Some assumptions are worse than others. For describing evolutionary processes, research on horizontal gene transfer indicates 'tangled-mazes-of-branches' is a better model than 'tree-of-life.' The WorldWideWeb replaced hierarchical trees of nouns at nodes with a web of links. Christopher Alexander said 'A city is not a tree.' Bureaucracies construct models as if everything is hierarchical, mirroring Conway's law: 'Organizations that design systems are constrained to produce designs that duplicate their own communication structures.' An authoritarian zoologist studied his pet dogs and identified two races *Der Dogg und Der Überdogg*. Economic models represent 'stylized facts – and there is no doubt they are stylized, though it is possible to question whether they are facts.' In statistics, to mathematize uncertainty, 'random errors' are defined as 'independent' (a prayer) and 'identically distributed' (a prank).

Models sanctified and celebrated by insiders can evolve into uncontested, lucrative, congealed monopolies/specialties/cartels/cults/disciplines – which in time, become self-centered and selfish, more and more about themselves, and less and less about their original substantive content. Local optimizing adds up to global pessimizing. Disciplines require hard-working true believers in local doctrines/assumptions that do not correspond to the truth. Revolutionary reforms are sometimes produced by those on the margin or outsiders. Two examples: invention of the Web by Tim Berners-Lee at CERN, the nuclear science research center; and when the magical thinking of classical economics and rational choice models was proved false by real world financial/economic disasters and by the deft empiricism of cognitive psychologists Amos Tversky and Daniel Kahneman.

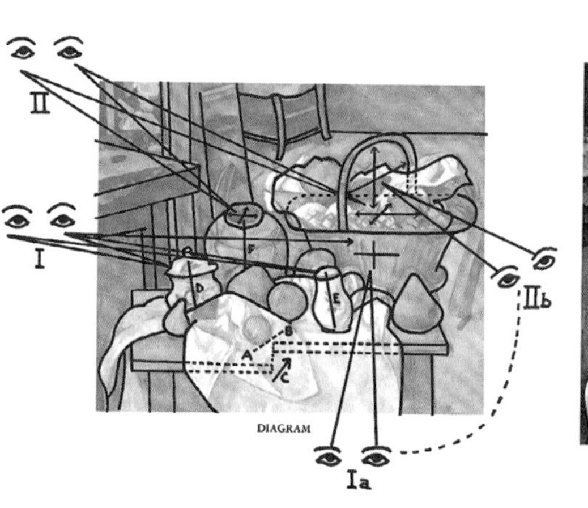

Cézanne deployed multiple local grids to compose local content, remodeling conventional one-eyed perspective models that had governed painting for centuries and photography for decades.

REMODELING MODELS/FORMATS/GRIDS

See with fresh eyes. Do not go lazy into default models, justified by "we've always done it this way"– words that end thought, censor deviations, block searches for alternatives. Nonetheless, many conventions and standards have got it right, or at least good enough, but fresh seeing and attempted remodeling can confirm their continuing righteousness.

Remodeling requires outsider comparison sets – insiders have exhausted their local fads and possibilities already. For example, compare an interface or data graphic side-by-side with a good map – since maps have solved inherent data display issues: color, content-located typography, massive data, diverse users. The remodeling proposals of this book:

maps inform and redesign data graphics/labels/interfaces

poetry/coding/math imply 2-dimensional graphical sentences

graphical sentences redesign conventional sentences and paragraphs

content-responsive typography replaces content-indifferent typography

webs of links and verbs destroy hierarchical trees of nouns

nameless statistical lives are taken as seriously as named lives

data analysis when the truth matters reforms conventional practices and teaching of statistics

credibility narratives replace false assumptions of standard statistical models of uncertainty

data paragraphs/small multiples/sparklines eliminate LittleDataGraphics

prevalence rates of 100s of specific statistical errors and frauds replace anecdotal horror stories

words annotate/model/explain links content-responsive arrays remodel image matrices

off-the-grid challenges on-the-grid ghostlier grids envision keener distinctions

lists escape conventional style-sheets documents and study hall improve standard presentations

thinking annotates the world text + image quilts replace book back-matter and references

fresh seeing challenges old conventions

signal replaces noise sombreros say *adios* to square hats

SOME MODELS ARE SUPREMELY BETTER THAN OTHERS:

A WEB OF LINKS REPLACES A HIERARCHY OF NOUNS,

CREATES A UNIVERSAL ARCHITECTURE FOR SHARING INFORMATION

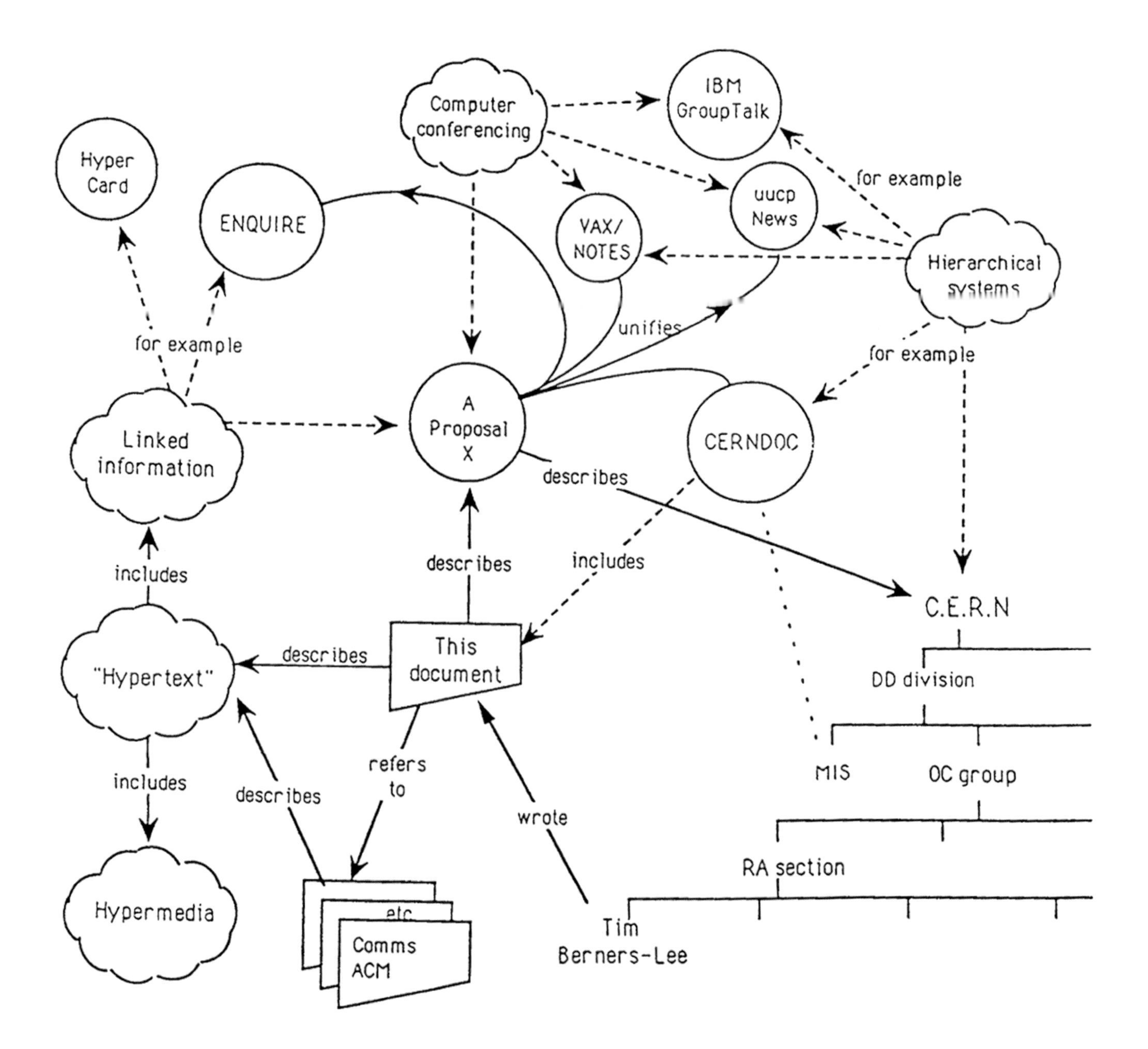

Tim Berners-Lee, 'Information Management: A Proposal,' CERN, March 1989. Tim Berners-Lee, *Weaving the Web* (1999) is an amazing, detailed account of putting together all the pieces of the WorldWideWeb while avoiding those who wanted to commercialize the web.

In the founding document of the web, Tim Berners-Lee pointed out that information management systems organize their content into grids of hierarchical trees with nouns at nodes. But information is not naturally organized into trees:

> **'The problem with trees**
>
> Many systems are organised hierarchically. The CERNDOC documentation system is an example, as is the Unix file system, and the VMS/HELP system. A tree has the practical advantage of giving every node a unique name. However, it does not allow the system to model the real world. For example, in a hierarchical HELP system such as VMS/HELP, one often gets to a leaf on a tree such as
>
> `HELP COMPILER SOURCE_FORMAT PRAGMAS DEFAULTS`
>
> only to find a reference to another leaf: Please see
>
> `HELP COMPILER COMMAND OPTIONS DEFAULTS PRAGMAS`
>
> and it is necessary to leave the system and re-enter it. What was needed was a link from one node to another, because in this case *the information was not naturally organised into a tree.'*

Direct, specific, side-by-side comparisons How did the broad words 'fails to model the real world' and 'the information was not naturally organised into a tree' turn into the web? As shown in the diagram, the proposed decentralized universal web of links is compared directly with the conventional brittle hierarchical model. A web is robust, unlike trees where a break separates the tree into unlinked pieces. If a web link breaks, the result is merely a 404 Not Found, indicating the server was reached but the desired page was not found on that server.

Universal and open: Information storage must not place restraints on the content

> The hope is to allow a pool of information to develop, grow, and evolve with the organisations and projects it describes. For this to be possible, **the method of storage must not place its own restraints on the information.**

Reasoning about real data, not toy data If CERN information – 1000s of research reports, millions of notes – can be web-modeled, any information can be.

The moral values of Tim Berners-Lee and CERN From the beginning, the web was conceived to be universal and open – not a business product. Tim Berners-Lee worked at a great scientific organization, an enterprise that seeks to create knowledge, not monetize it. (CERN did not patent the Higgs Boson.) Four years after the invention of the WorldWideWeb, CERN gifted full rights to the world: no licensing, no lawyering up, no copyrights, no patents, no marketing, no royalties, no advertising against the stored information – not for private enrichment, but for public good. Then came the sharks.

CONCRETE, SPECIFIC COMPARISONS IN MODELING AND REMODELING

"I do not paint things, I paint only the differences between things" HENRI MATISSE

Ken Kocienda, former principal engineer of Apple iPhone software:

"Exactly *how* we collaborated mattered, and it reduced to a basic idea: We showed demos to each other. Every major feature on the iPhone started as a demonstration, and for a demo to be useful to us, it had to be *concrete and specific.* We needed concrete and specific demos to guide our work, since even an unsophisticated idea is hard to discuss constructively. Here's an example:

Think of a cute puppy. Picture one in your mind. Close your eyes if you need to. Make the image as detailed as you can. Take a moment. A cute puppy. Got one? I do too, and I did well. In fact, I think my puppy is cuter than yours.

Consider the scenario. Two people have imagined two cute puppies. I assert mine is cuter. What do we do now? Do we have a *cuteness* argument? How can we? We have nothing to go on. Do I try to describe the puppy in my mind and attempt to sway you that my vision of a golden retriever puppy is superlatively cute – because *everyone knows* that golden retrievers are the cutest of all dog breeds – and therefore, my conjured mental picture is unbeatably cute. Do you try to make a sketch on a whiteboard of the puppy you're thinking of but then apologize because you're a lousy artist, so I'll just have to take your word for how cute your puppy really is in your mind? Let's say you're my manager. What do you do now . . . pull rank? The scenario is ridiculous. There's no way to resolve this conflict. Without a concrete and specific example of a cute puppy, there's no way to make progress.

Now, I can make this easier. Here are pictures of two cute puppies. Now we can talk about the merits of these options. I can make my case for the cuteness of the golden retriever on the left. You might favor the lovable bulldog and attempt to persuade me that the dog-smiley happy face and single flopped-over ear make it cuter. I might argue back, pointing out the extraordinarily cute way the retriever's paws are buried in the not-so-tall grass. If we weren't satisfied with these two choices, we could search the web for countless others.

Concrete and specific examples make the difference between a discussion that is difficult and perhaps *impossible* to have, compared to one that feels like child's play. At Apple, we built our work on this basic fact. Demos made us react, and the reactions were essential. Direct feedback on one demo provided the impetus to transform it into the next."

Ken Kocienda, *Creative Selection: Inside Apple's Design Process During the Golden Age of Steve Jobs*, 2018, 153-155, edited.

"Demos were the catalyst for creative decisions, and we found that the sooner we started making creative decisions – whether we should have big keys with easy-to-tap targets or small keys coupled with software assistance – the more time to refine those decisions, to backtrack if needed, to forge ahead if possible. Making demos was the core of the process of taking an idea from the intangible to the tangible.

Making demos is hard. It involves overcoming apprehensions about committing time and effort to an idea that you aren't sure is right. At Apple, we then had to expose that idea and demo to the scrutiny of sharp-eyed colleagues who were never afraid to level pointed criticism. Most demos – almost all of them – failed in the absolute, dead-end sense of the word. This prospect of failure can make it tough to sit down, focus, make a demo at all. Getting another cup of coffee sounds better, and then after the communal break, the whiteboard may beckon, and the group might veer off into a gab session. We rarely had brainstorming sessions. I recall only a few times in my Apple career I stood around to rough out big plans at a whiteboard."

HONESTY IN COMMUNICATION

"Scott Forstall, Senior Vice President of Software, kept his job as final demo gatekeeper for Steve because he advanced work only when it was of sufficiently high quality. Scott knew that he couldn't bring third-rate demos, with a claim that his team had toiled away on it assiduously, hoping to weasel his way to a CEO buy-off. That would have been a good way to trigger Steve's temper and a good way for Scott to lose his role as second-to-last decider. Scott also knew that he couldn't corrupt the demo procedure by trotting out the equivalent of one fast thoroughbred mixed in with mediocrities, expecting Steve would be happy he could pick the winner. Steve could see right through such ploys.

For Steve, it was crucial to give honest feedback in these demo meetings, and he could be brutally straight-forward when it came to voicing opinions about projects. It's true that Steve did go overboard and act in ways that tried the patience of those around him – another reason why these meetings with him were kept small. It was necessary to have a thick skin when showing work to Steve. Yet, in my experience at Apple, the mix of honesty and decisiveness in CEO demos with Steve was essential to the quality of the final work.

Demo reviews were also part of Steve's effort to model the product development behaviors he wanted us to use when he couldn't be present. The whole software organization kept meetings and teams small to maintain efficiency and to reinforce the principle of doing the most with the least. Steve's constant demand to see a succession of demos spawned numerous other demos with their own presenters and deciders, and each was built on a frank exchange of opinions about what was good – and what wasn't."

Ken Kocienda, *Creative Selection: Inside Apple's Design Process During the Golden Age of Steve Jobs*, 2018, 153-155, edited.

ELABORATE SPACING MAKES CODE INTELLIGIBLE

```
function Knuth_Morris_Pratt(const target, pattern : PChar; const lTarget, lPattern : integer) : integer;
var
  step  : array[0...255] of integer;                // failure table
  i,                                                // main loop
  j     : integer;
begin
  result := -1;
  if lTarget * lPattern = 0 then exit;
  i := 0;
  j := -1;
  step[0] := -1;
                                                    // preprocessing the table
  repeat
    if (j = -1) or (pattern[i] = pattern[j]) then
    begin
      inc(i);
      inc(j);
      if pattern[j] = pattern[i] then step[i] := step[j] else step[i] := j;
    end else j := step[j];
  until i = lPattern - 1;
  j := -1;
  i := 0;
                                                    // search main loop
  while i < lTarget do
  begin
    if (j=-1) or (pattern[j] = target[i]) then      // comparison loop
    begin
      inc(i);
      inc(j);
      if j >= lPattern then                         // Match found
      begin
        result := i-j+1;
        exit;
      end;
    end else j := step[j];                          // skips the value found in the table
  end; // while
end; // Knuth_Morris_Pratt
```

Computer code resides in the plane, with content-driven spaces within and between lines. And the left margin indents lines to match their content. Like poetry and math, code is not hyphenated at squared-off right margins. Above, the Knuth-Morris-Pratt string matching algorithm is 37 lines and 75% empty spaces. For coders, spaces create fluency; for machines, spaces mean nothing and are skipped over by compilers. Below, minified KMP code without spacing in one long line. This executes properly, but is very difficult for humans to read:

```
function Knuth_Morris_Pratt(const target, pattern : PChar; const lTarget, lPattern : integer) : integer; var
step : array[0...255] of integer; i, j : integer; begin result := -1; if lTarget * lPattern = 0 then exit;
i := 0; j := -1; step[0] := -1; repeat if (j = -1) or (pattern[i] = pattern[j]) then begin inc(i); inc(j);
if pattern[j] = pattern[i] then step[i] := step[j] else step[i] := j; end else j := step[j]; until i =
lPattern - 1; j := -1; i := 0; while i < lTarget do begin if (j=-1) or (pattern[j] = target[i]) then begin
inc(i); inc(j); if j >= lPattern then begin result := i-j+1; exit; end; end else j := step[j]; end; end;
```

High-resolution display screens benefit code text-editors with spaces clarifying content, and multiple columns showing contextual information, including vertically compressed code.

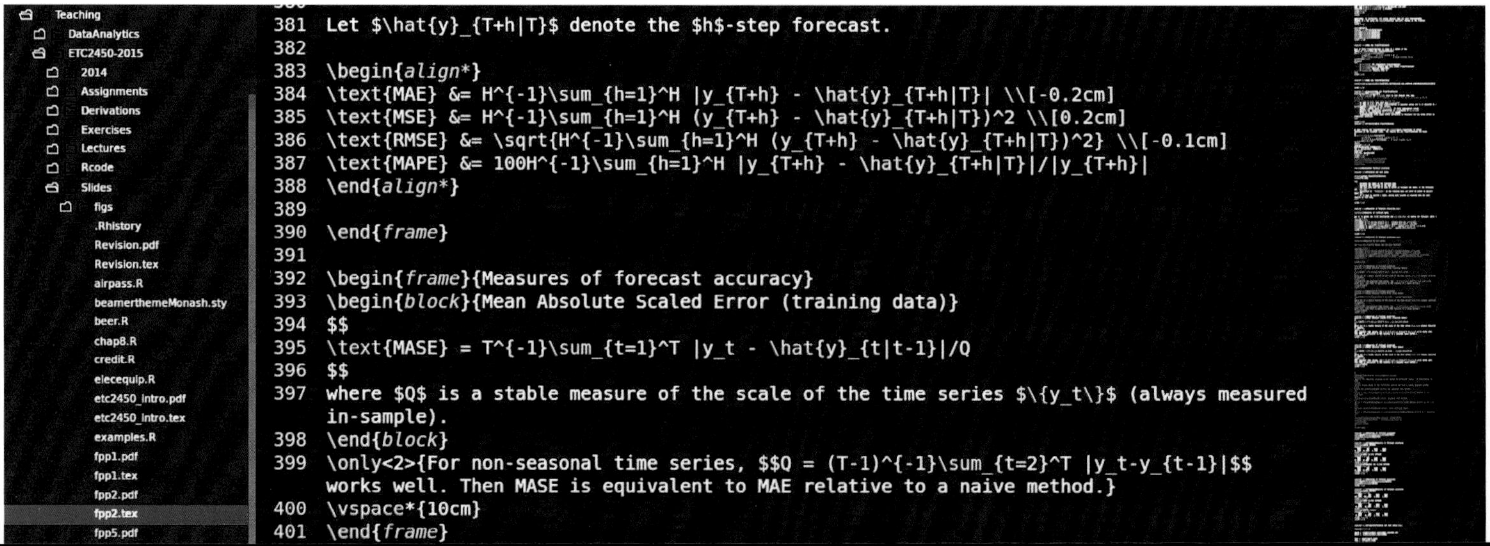

Teaching
DataAnalytics
ETC2450-2015
2014
Assignments
Derivations
Exercises
Lectures
Rcode
Slides
figs
.Rhistory
Revision.pdf
Revision.tex
airpass.R
beamerthemeMonash.sty
beer.R
chap8.R
credit.R
elecequip.R
etc2450_intro.pdf
etc2450_intro.tex
examples.R
fpp1.pdf
fpp1.tex
fpp2.pdf
fpp2.tex
fpp5.pdf

```
381 Let $\hat{y}_{T+h|T}$ denote the $h$-step forecast.
382
383 \begin{align*}
384 \text{MAE} &= H^{-1}\sum_{h=1}^H |y_{T+h} - \hat{y}_{T+h|T}| \\[-0.2cm]
385 \text{MSE} &= H^{-1}\sum_{h=1}^H (y_{T+h} - \hat{y}_{T+h|T})^2 \\[0.2cm]
386 \text{RMSE} &= \sqrt{H^{-1}\sum_{h=1}^H (y_{T+h} - \hat{y}_{T+h|T})^2} \\[-0.1cm]
387 \text{MAPE} &= 100H^{-1}\sum_{h=1}^H |y_{T+h} - \hat{y}_{T+h|T}|/|y_{T+h}|
388 \end{align*}
389
390 \end{frame}
391
392 \begin{frame}{Measures of forecast accuracy}
393 \begin{block}{Mean Absolute Scaled Error (training data)}
394 $$
395 \text{MASE} = T^{-1}\sum_{t=1}^T |y_t - \hat{y}_{t|t-1}|/Q
396 $$
397 where $Q$ is a stable measure of the scale of the time series $\{y_t\}$ (always measured
    in-sample).
398 \end{block}
399 \only<2>{For non-seasonal time series, $$Q = (T-1)^{-1}\sum_{t=2}^T |y_t-y_{t-1}|$$
    works well. Then MASE is equivalent to MAE relative to a naive method.}
400 \vspace*{10cm}
401 \end{frame}
```

SPACES AND LINEBREAKS CREATE POETIC MEANING

More than meter, more than rhyme,
more than images or alliteration or figurative language,
line is what distinguishes our experience of poetry as poetry,
rather than some other kind of writing. JAMES LONGENBACK

Historically, line has been the characteristic unit distinguishing poetry from prose;
it is the most sensitive barometer of breath-units in which poetry is voiced. The very shortest
way of composing a line makes a single word constitute a line; the very longest manner
of composition invents a line that spills over into turnovers, or suspends from its right margin
an appended short line, what Hopkins called an 'outride.' When a poet ceases to write short lines
and starts to write long lines, that change is a breaking of style almost more consequential,
in its implications, than any other. HELEN VENDLER

Whether end-stopped or enjambed, the line in a poem moves horizontally,
the rhythm and sense also drive it vertically, and the meaning continues
to accrue as the poem develops and unfolds. EDWARD HIRSCH

Read aloud John Ashbery's poem *Syringa.* Lines 1-2 brilliantly describe the one-time experience of seeing with fresh eyes without patterned expectations:

1 The seasons are no longer what they once were,
2 But it is the nature of things to be seen only once,
3 As they happen along, bumping into other things, getting along
4 Somehow.

Line 3 ends in a linebreak without punctuation – on the edge of line 4 – then upon arrival, a single stark word and full-stop: Somehow.
Line 3 is an 'enjambed line,' a line spilling over to the next line.

5 of the first 8 lines in T.S. Eliot's *The Waste Land* are enjambed:

April is the cruellest month, breeding
Lilacs out of the dead land, mixing
Memory and desire, stirring
Dull roots with spring rain.
Winter kept us warm, covering
Earth in forgetful snow, feeding
A little life with dried tubers.

Linebreaks without enjambment tranquilize the 8 lines:

April is the cruellest month,
Breeding lilacs out of the dead land,
Mixing memory and desire,
Stirring dull roots with spring rain.
Winter kept us warm,
Covering Earth in forgetful snow,
Feeding a little life with dried tubers.

Does central axis typography assist reading poetry aloud?

April is the cruellest month, breeding
Lilacs out of the dead land, mixing
Memory and desire, stirring
Dull roots with spring rain.
Winter kept us warm, covering
Earth in forgetful snow, feeding
A little life with dried tubers.

WHEN MODELS BREAK DOWN: AUTHORITARIAN 3D-GRID MICRO-MANAGES DISHWASHER LOADING, REMODELS DOMESTIC TRANQUILITY

This bizarre decision tree for placing silverware in a dishwasher challenges the computational powers of a chess champion. These dishwasher loading tips create an interface that mutates into a stress test for domestic relationships:

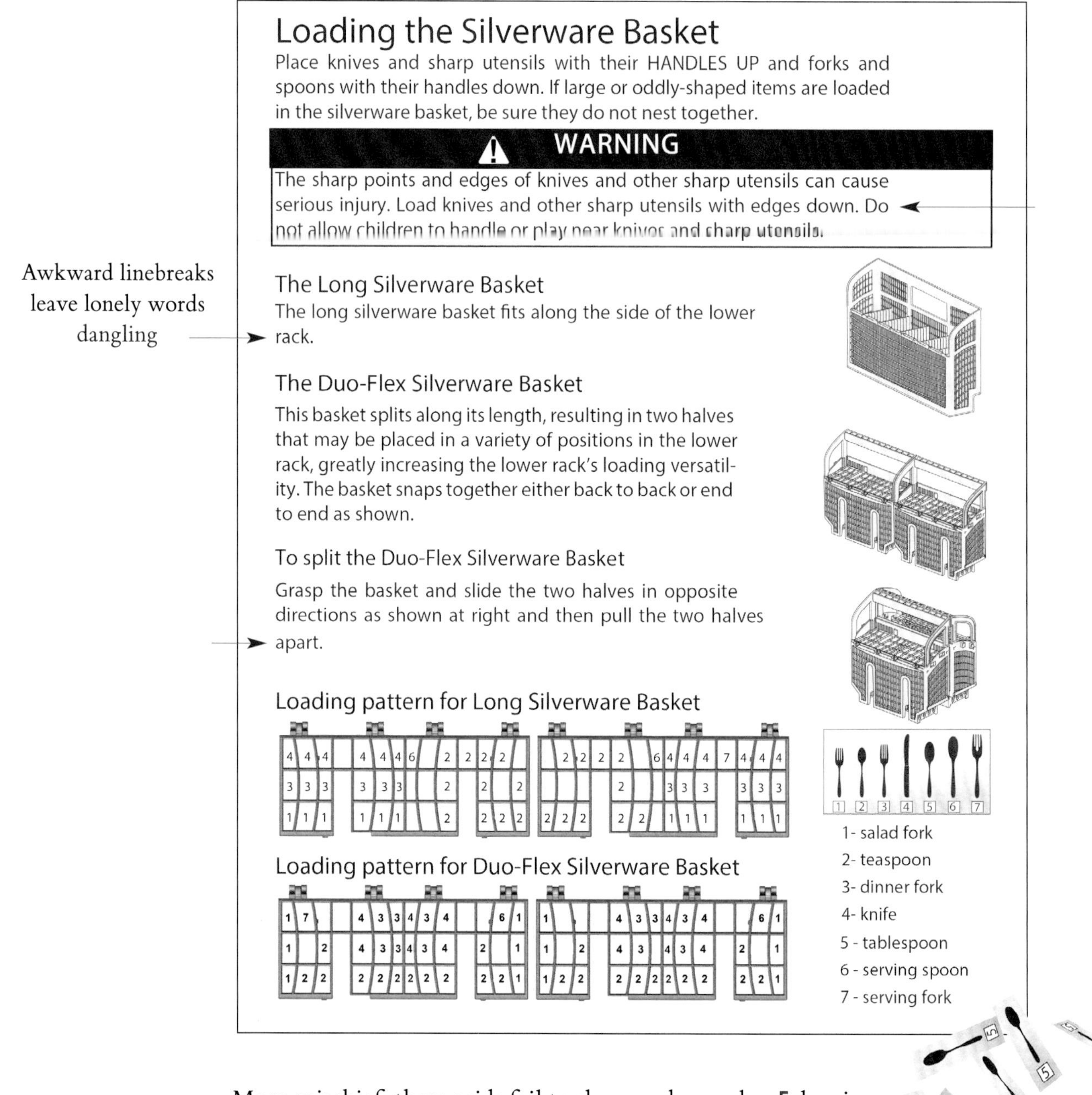
Awkward linebreaks leave lonely words dangling

Loading the Silverware Basket

Place knives and sharp utensils with their HANDLES UP and forks and spoons with their handles down. If large or oddly-shaped items are loaded in the silverware basket, be sure they do not nest together.

WARNING

The sharp points and edges of knives and other sharp utensils can cause serious injury. Load knives and other sharp utensils with edges down. Do not allow children to handle or play near knives and sharp utensils.

The Long Silverware Basket

The long silverware basket fits along the side of the lower rack.

The Duo-Flex Silverware Basket

This basket splits along its length, resulting in two halves that may be placed in a variety of positions in the lower rack, greatly increasing the lower rack's loading versatility. The basket snaps together either back to back or end to end as shown.

To split the Duo-Flex Silverware Basket

Grasp the basket and slide the two halves in opposite directions as shown at right and then pull the two halves apart.

Loading pattern for Long Silverware Basket

Loading pattern for Duo-Flex Silverware Basket

1- salad fork
2- teaspoon
3- dinner fork
4- knife
5 - tablespoon
6 - serving spoon
7 - serving fork

More mischief: these grids fail to show code number **5**, leaving users with an annoying handful of #☠☆¡!! tablespoons.

5-tablespoon

THOUGHT BUBBLE TYPOGRAPHY

A LOCAL GRID WITH METAPHORICAL MEANING: COMPASSION IN MEDICAL CARE

Cleveland Clinic's video *Empathy: The Human Connection to Patient Care* teaches and prompts medical caregivers to think about the circumstances and thoughts of their patients and colleagues. In brief silent scenes, the only words are graphical sentences that move with the patients and describe what's going on in the minds of patients and staff. Widely viewed in medical classes, this video poses thoughtful and rigorous questions about compassion:

"Can we see through someone else's eyes, stand in their shoes, hear what they hear?

What if you could see a thought bubble above every person's head,
that shows you what they are thinking and feeling as you see them?

If so, would you treat them differently?"

Compassion/empathy ideals are exemplified by a thought-bubble metaphor, an invisible and dynamic local grid. Thought-words in graphical sentences move along, each person living those words (not superimposed over people, a quiet signal of respect).

content locates typography *space, location, linebreaks create and clarify meaning*

ELECTRONIC HEALTH RECORDS SEIZE OWNERSHIP OF MEDICAL PATIENT INFORMATION, MEDICAL CENTER BUSINESS PLANS = OWN THE DATA, OWN THE PATIENT

Medical patient ending appointment: "Doctor, today I learned more about your EHR system than about my heart."
Cardiologist, unable to find recent test results, ordered duplicate tests, says: "Me too."

Electronic Health Records violate Tim Berners-Lee's fundamental principle for data models: "The hope is to allow a pool of information to develop, grow, and evolve. For this to be possible, **the method of storage must not place its own restraints on the information.**" Local EHRs, however, seize ownership by copyrighting patient content. Every day, logging into an EHR, millions of patients and staff must accept a bullying gag order that looks much like this parody:

Prior to proceeding, you must agree to every word below governing use of The System

All content included in the **Foundation University Healthcare Systems**, including, but not at all limited to, colors, words, photos, graphs, icons, buttons, graphics, images, videos, feature-length films, numbers (finite, rational, irrational, imaginary, real, troubled, prime), punctuation (including the Palatino Linotype interrobang), artworks, proper nouns, logos, trademarks, data (the "Content"), in all and any forms including compilations, are protected by all laws and conventions. Except as set forth no where, direct indirect reproduction (forget screenshots) of "Content" or The System, by any means, are prohibited without explicit consent of Interpol and **Foundation University Healthcare Systems**. ACCEPT! DECLINE

Don't even think of clicking on DECLINE, for you will arrive at The Shadowed Box threatening to **disable** your medical record – a mean nasty threat to frazzled patients. This is not a parody. Medical centers pitch empathy in their marketing, but intimidate patients to sign gag orders seizing ownership of all medical patient records.

If you fail to agree with **FUHS** Terms and Conditions, your medical record **will be disabled** and you will need to contact **Customer Services** to access your medical record.

GO BACK! CONTINUE?

Patients have enough problems, and will give up on **Customer Services.** *Inconvenient opt-out is* inherent to software business models. Also medical patients appear to be redefined by the EHR as "users" and "customers". Do medical patients thereby lose their unique legal rights?

In university medical centers, EHR systems may violate university norms: *freedom of speech and inquiry, civility and respect for others, even anti-plagiarism rules.* Gag orders also *stifle and block interface research* by prohibiting screenshots of any element of the EHR interface in research and professional meetings. Yet EHR interfaces do medical center command and control – and gag orders impede assessments by experts and researchers of this crucial interface. Why do university medical centers agree to this?

Proprietary EHRs are fundamental to U.S. medical center business models: own the data, own the patient, monetize everything, take over local competitors, make referrals to doctors within the System – and, in the U.S., charge monopoly prices, do predatory surprise/billing followed up by automated debt collectors that may bankrupt patients. Vendor capture of customers engages proven business models of proprietary software and Tony Soprano's Waste Management. EHR Systems are governed by medical center suits, Senior Vice Presidents for Finance and Marketing. In the U.S., the suits maximize profits and report to commercial interests and investment bankers. The result is vast transfers of money from the sick to the rich. Financial interests directly conflict with the interests of patients, who seek to remain alive and healthy, and not bankrupted.

Electronic Health Records are *inherent to patient care:* the medical staff communicate with patients, adjust prescriptions, enter orders/notes, make referrals (inside to the System), and guide medical decisions. Despite vast public investment in EHRs, much of patient data is still communicated by fax or hand-carried by patients. Busy clinics receive thousands of fax pages daily. Hint to patients: get copies of all test reports, bring them to every medical appointment for you and your family.

In EHR encounters, are there grounds for a Medicare Quality of Care Grievance and a Plan Grievance Report (EHR rudeness)? Medicare rules for patient grievance reports (why no class-actions?) include:

> "Examples of problems that are typically dealt with through the Quality of Care Grievance process: Duplicate tests, with possible side effects and adverse reactions. . ."

> "Examples of problems that are typically dealt with through the Plan Grievance process: Disrespectful or rude behavior by doctors, nurses or other plan, clinic, or hospital staff. Long waiting times. Difficult to make appointments."

In medical care, everything within 50 meters of patients must follow fussy and detailed regulatory, industrial, professional standards. *Where is the evidence that EHRs are safe and effective, that benefits exceed harms to patients and medical staff?* To not answer these questions, each EHR installation has conducted one of the worst clinical trials ever: patients and staff are enrolled without consent in a vast unrandomized uncontrolled experiment, without measuring outcomes/harms/benefits, and with no plans for stopping the trial in event of excessive harms. And where was the Human Subjects Safety Review Board?

EHR problems are well-documented, and poignantly by Atul Gawande, *Why Doctors Hate Their Computers.* Eventually, perhaps, patients will some day own their medical records, despite resistance by U.S. hospital trade associations (which include University Medical Centers) seeking to own all patient records and to enforce their business models.

REMODELING MEDICAL PATIENT HEALTH RECORDS:
WHY A MEDICAL PATIENT'S HEALTH RECORD MUST BELONG TO THE PATIENT (BY ERIC TOPOL)

"It's your body

You paid for it

It is worth more than any other type of data

It's widely sold, stolen, hacked. And you don't know it

It's full of mistakes, that keep getting copied and pasted, that you can't edit

You will be generating more of it, but it's homeless

Your medical privacy is precious

The only way it can be made secure is to be decentralized

It is legally owned by doctors and hospitals

Hospitals won't or can't share your data ('information blocking')

Your doctor (>65%) won't give you copies of your office notes

You are far more apt to share your data than your doctor

You'd like to share it for medical research, but you can't get it

You have seen many providers in your life, but no health system/ insurer has all your data

No one (in US) has all their medical data from birth throughout their life

Your EHR was designed to maximize billing, not to help your health

You are more engaged, have better outcomes when you have your data

Doctors with full access to patient records look at them routinely

It requires comprehensive, continuous, seamless updating

Your access or 'control' of your data is not adequate

~10% of medical scans are unnecessarily duplicated due to inaccessibility of prior scans

You can handle the truth

You need to own your data; it should be a civil right

It could save your life "

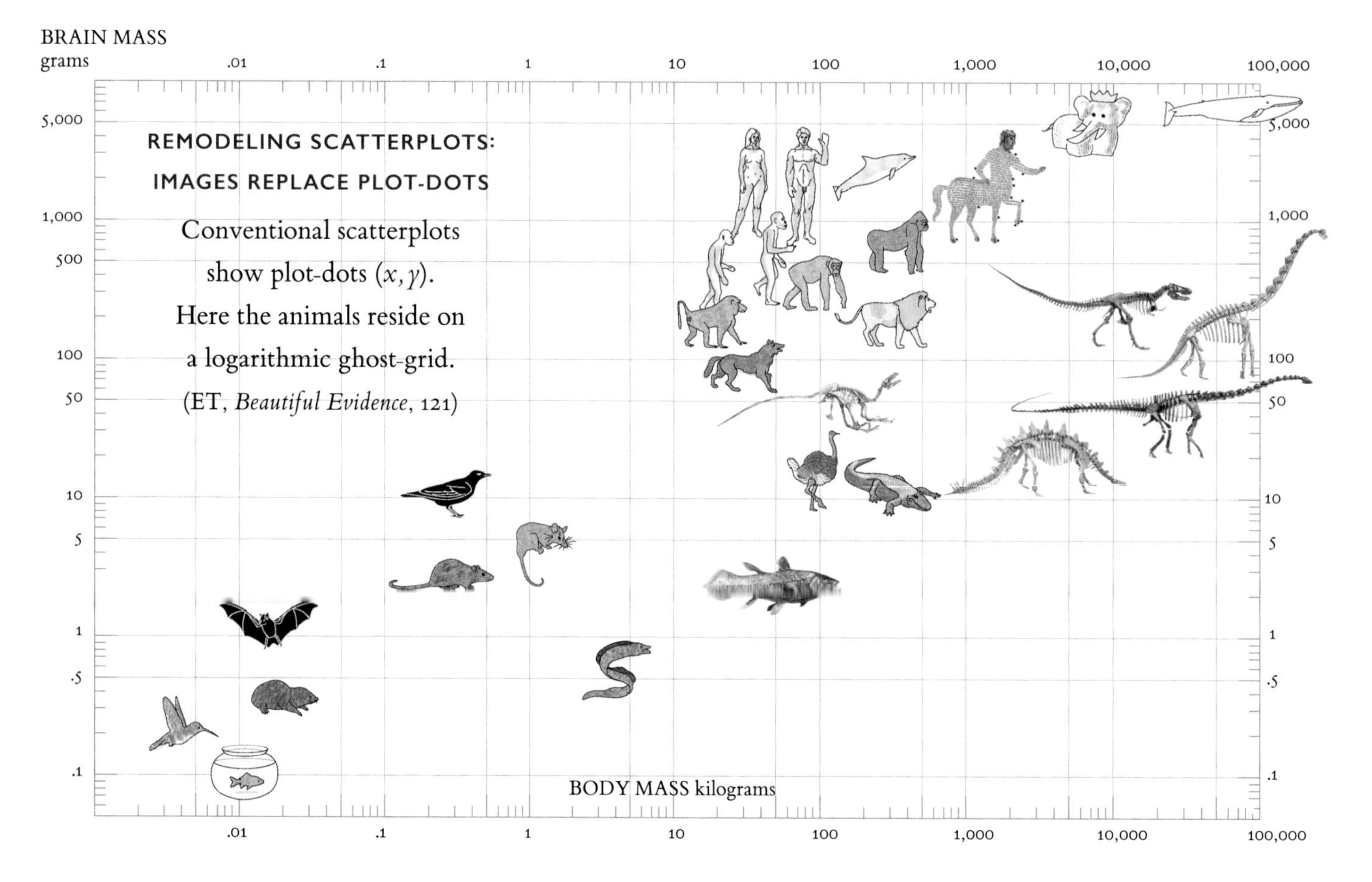

REMODELING SCATTERPLOTS: IMAGES REPLACE PLOT-DOTS

Conventional scatterplots show plot-dots (x, y). Here the animals reside on a logarithmic ghost-grid. (ET, *Beautiful Evidence*, 121)

REMODELING SPREADSHEETS: BIG DATA SPARKLINES REPLACE THE SINGLE LONELY NUMBER IN EACH CELL

In the video game PAC-MAN, values of each frame (60 fps) by active memory address were recorded for 10 seconds, yielding 264 sparklines. In an intense display by Michael Fogleman, each sparkline represents many numbers in each cell. The 264 sparklines are arrayed on a invisible grid.

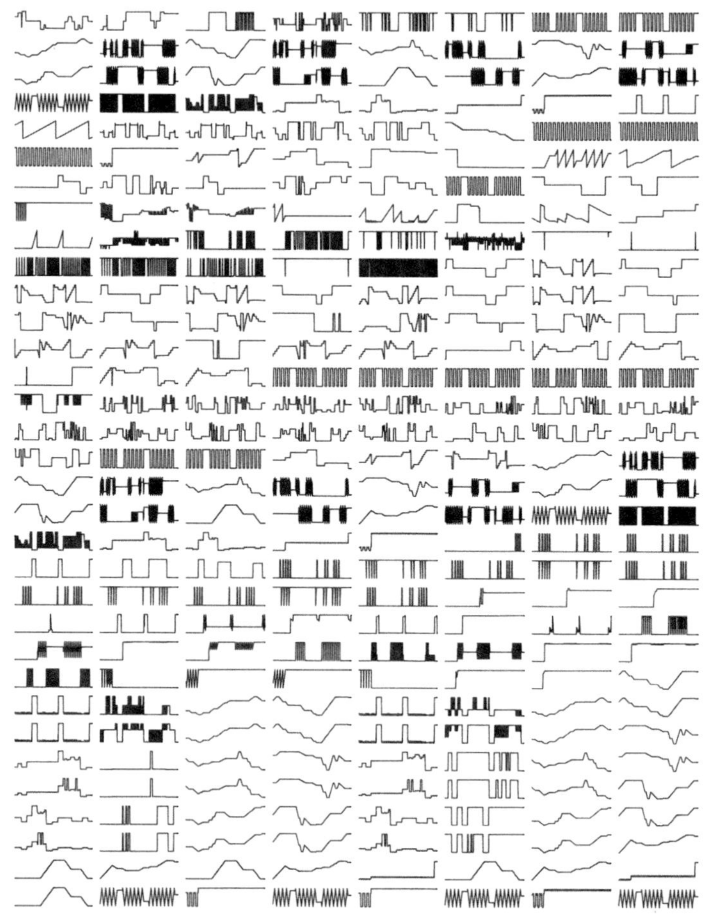

MAYA LIN'S WOMEN'S TABLE AT YALE: A SPIRAL DATA MODEL EXPRESSES A FAR-REACHING MESSAGE

Maya Lin: "It's open-ended. Commemorating women at Yale, we have a beginning, but certainly, it's ongoing. So I thought of a spiral. I was looking at one of Edward Tufte's books, *Envisioning Information.* He had a beautiful spiral of numbers. He's a professor at Yale. I called him up, and both Professor Tufte and his wife Inge Druckrey, who's in graphic design, were unbelievably helpful in helping me come up with the spiral numbers, how to lay it out, everything. And it's a spiral starting with many zeros signifying there were no women at Yale for a very, very long time."

spiral graphic of the periodic table of elements from *Envisioning Information*

number of women students by year (ET mock-up)

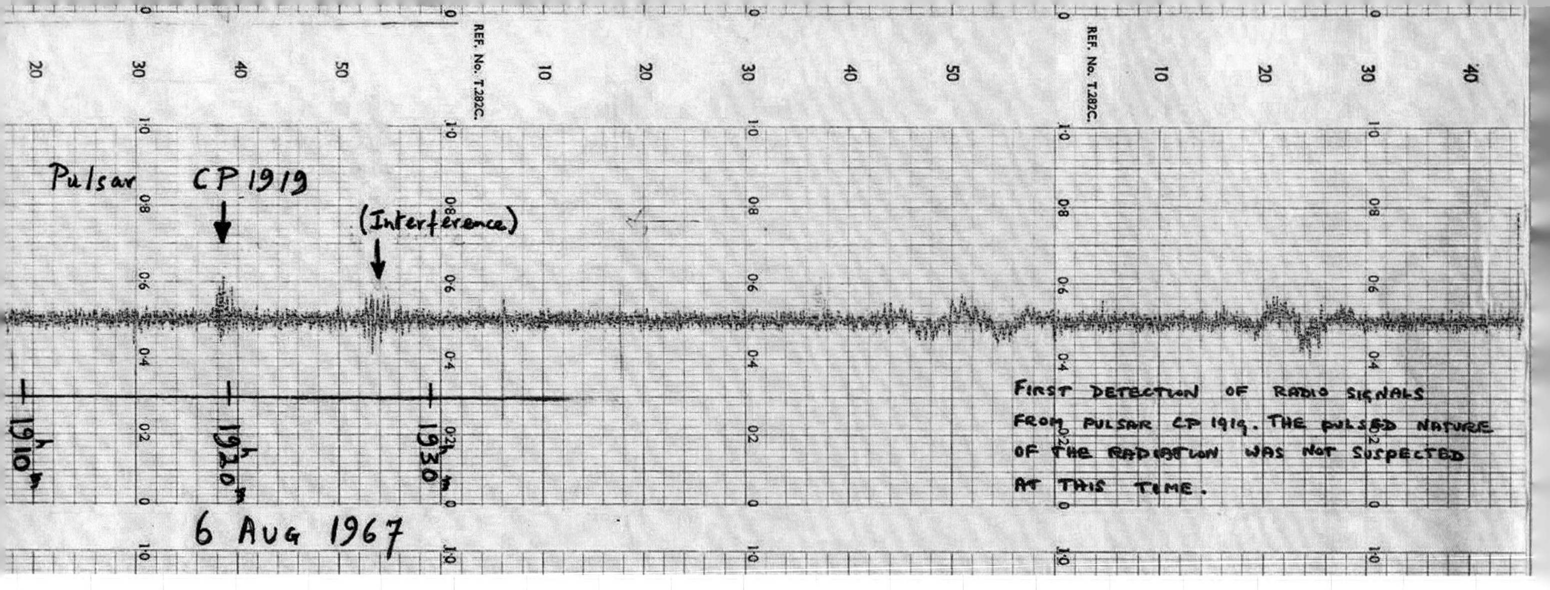

Jocelyn Bell Burnell recognized evidence of pulsars (rotating neutron stars) and annotated her data traces at the Cambridge 4 Acre Array radio telescope. The grid paper provides exactitude, clarity, permanence.

GHOSTLIER GRIDS ALLOW KEENER DISTINCTIONS

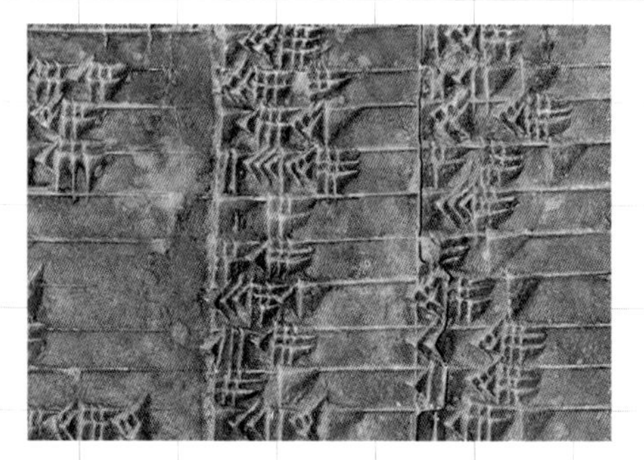

A delicate grid in clay showing a table of right-triangle perfect squares 1800 BCE

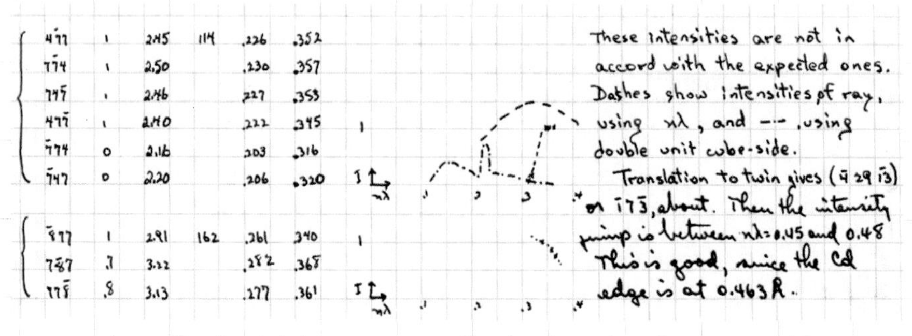

Linus Pauling's laboratory notebooks show hand-written tables, text, graphics on a *ghost-grid,* with coordinates for data and axis I ↑ nλ located near (0,0) 1922

CONTENT-HOSTILE GRIDS GENERATE OPTICAL NOISE, REDUCE SIGNAL QUALITY

Image matrix is only 42% images, with 58% non-content spaced-out frame stuff, over-large checkmarks, tiny word labels.

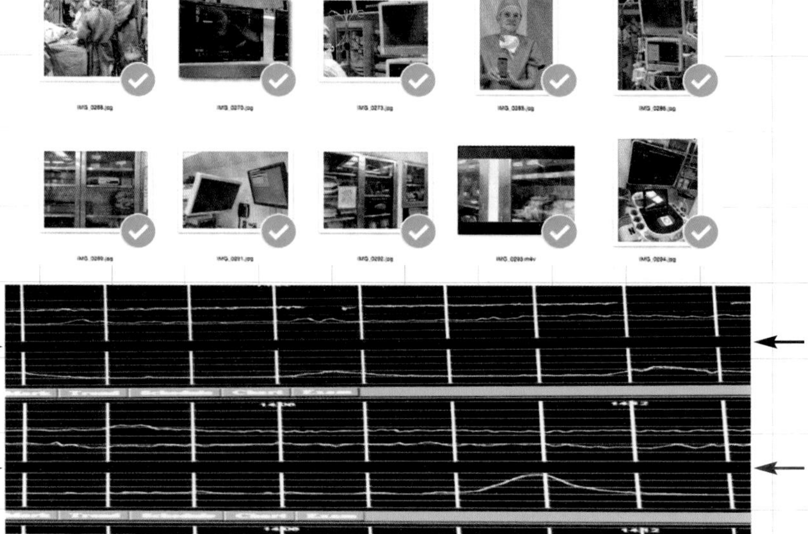

This real-time medical interface increases noise, reduces signal. *The strongest visual elements are not data,* but instead cyan grid bars and buzzy verticals. Gaps in → white verticals create optical vibration and illusory horizontal stripes.

1 CENTIMETER GHOST GRID (METRIC)

GHOSTLIER GRIDS ALLOW KEENER DISTINCTIONS

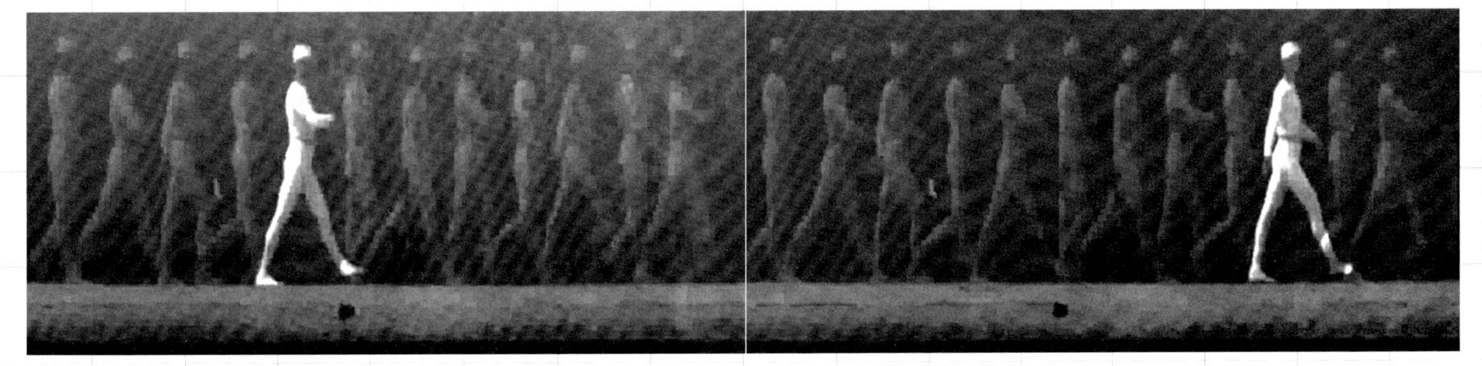

E. J. Marey's *ghosted substrate grid* in early films of animal and human movement ~1880

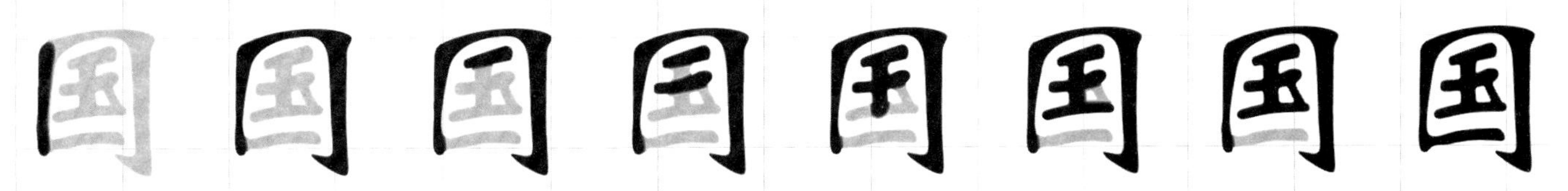

A space of spaces: 8 ghosts of the completed character provide sequential instructions to guide students through the 8 required strokes in this Chinese brushstroke lesson book. The general rule for stroke order is 'start upper left, work downward.' Calligraphically-correct stroke order varies for different versions of Chinese, each enforced by local stroke-order police.

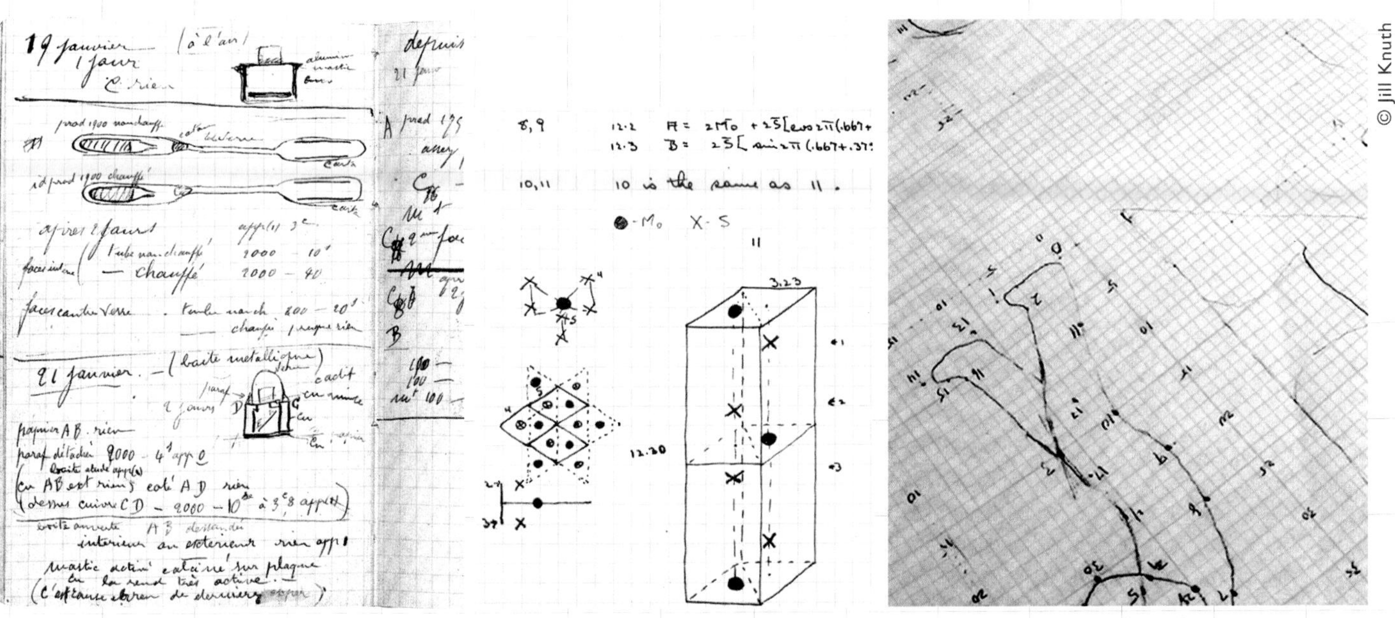

© Jill Knuth

Marie Curie 1899-1902 | Linus Pauling 1922 | Donald Knuth 2019

Some composition and lab notebooks use over-active thick grid lines that argue with content. Better to use ghost-grids, as in these scientific notebooks of Marie Curie (2 Nobel Prizes in science). Linus Pauling (Nobel Prize in science), and Donald Knuth. They recorded their work on grid lines clear enough to maintain approximate horizontal/vertical alignments among 3D drawings of physics equipment, atomic structures, text, tables, graphics – but ghostly enough to avoid optical clutter.

GHOSTED GLOBAL DATA SUBSTRATE = CONTEXT FOR 61 COUNTRIES

This small multiple shows daily deaths from covid-19 for 61 countries. Ghosted trajectories, similar to thin contour lines on topographic maps, locate and scale each country in the context of all countries.

John Burn-Murdoch at the newspaper *Financial Times* analyzed data from the European Centre for Disease Prevention and Control. The worldwide language of pandemics is data, as the worldwide language of money is data. But many news outlets are local, named after cities or countries, and write about locals in local languages, favor anecdotal stories, naively believe data is boring, inhuman, lacking in compassion. The *Financial Times* is global, collects data worldwide, and produced these superb graphics.

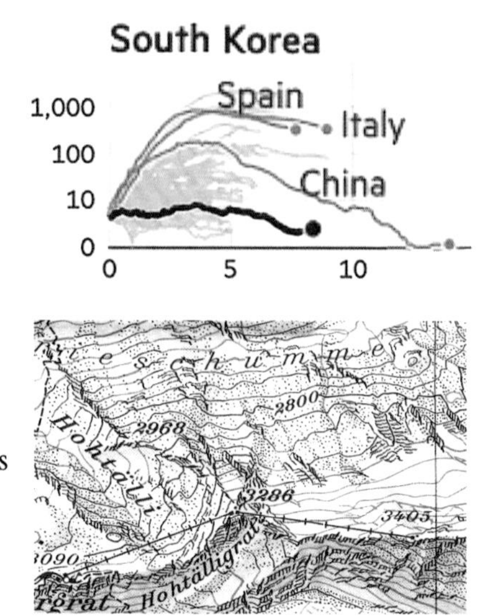

Daily death tolls have now peaked in many countries

Daily deaths with coronavirus (7-day rolling average), by number of weeks since 3 daily deaths first recorded

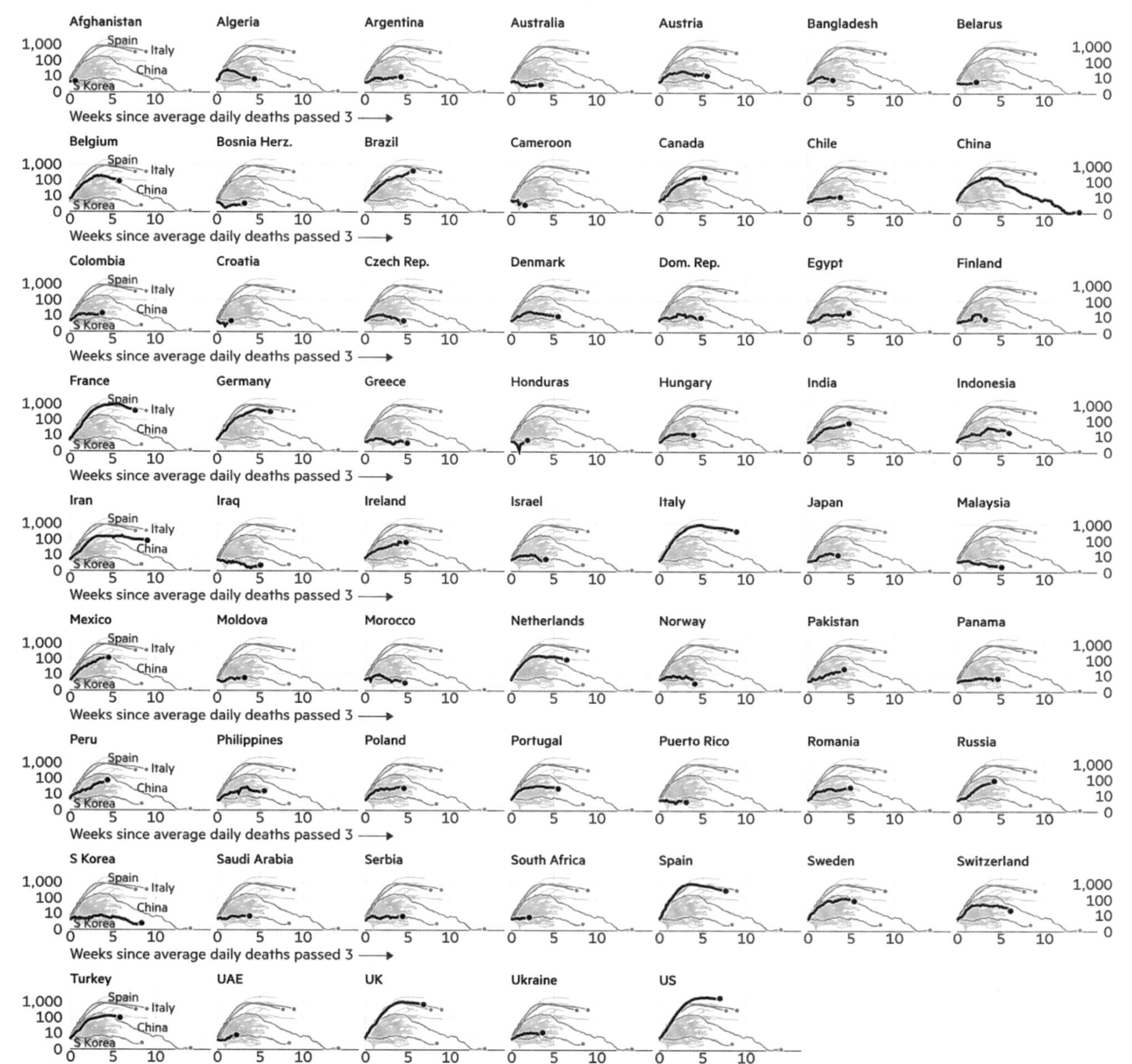

FT graphic: John Burn-Murdoch / @jburnmurdoch   Source: FT analysis of European Centre for Disease Prevention and Control; FT research. Data updated May 02, 22:18 BST   © FT

SONATA II

19

Grave

f espr.

meno f

più f

sempre f espr.

sfz *molto espr.*

rit. *p* *molto tranq. sempre p e sotto voce*

in der II. Lage bleiben

*) Triller mit „d" beginnen

13900 (3)

MUSICAL SCORE BECOMES A SUBSTRATE GRID FOR ANNOTATION

Yehudi Menuhin, a great violinist, marked-up a score for Bach's *Sonata No. 2 for Solo Violin.* Penciled annotations show real-time performance strategies. To outsiders, insider mark-ups appear chaotic and cryptic, but these personal annotations are for Menuhin's eyes, the only eyes that matter. All can learn from this useful workaday grid strategy: *a relevant and intense data layer serves as a coherent substrate scaffold upon which to overlay additional information.* Maps do this all day long.

DAVID HOCKNEY PEARBLOSSOM HWY. 11–18TH APRIL 1986, #2 1.82 X 2.72M

SPACES OF SPACES

David Hockney on conventional one-point perspective grids: 'Photography is all right, if you don't mind looking at the world from the point of view of a paralyzed Cyclops – for a split second.' Hockney's collage assembly of *750 photographs* each with Cézanne-like perspectives from varying angles and positions replaces one-eyed linear perspective. 'Every photograph here is taken close to something because cameras do push you away. I was trying to pull you in.'

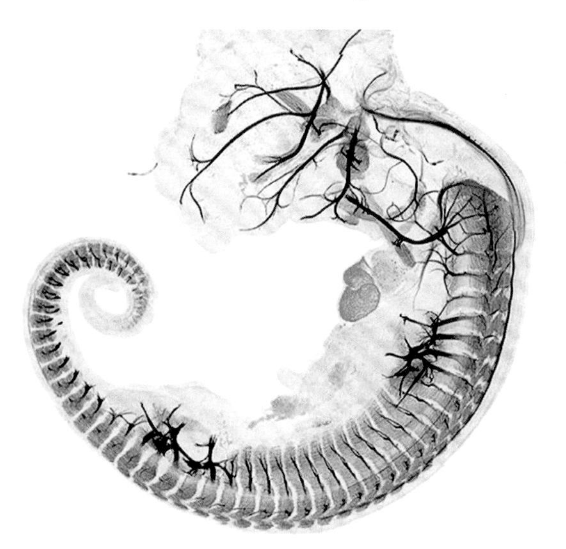

Like David Hockney's collage, this photo collage combines '12,000 individual images of the muscles and nerves of a Madagascar ground gecko embryo after 12 days of incubation in the egg. These 12,000 images were made by a Zeiss LSM 880 confocal microscope. *Confocal* means the microscope draws light from only a small part of the specimen.'

BHART-ANJAN BHULLAR
AND DANIEL SMITH-PAREDES 2018

© 2019 Jessie T. Pettway/Artists Rights Society (ARS, NY)

© 2019 Loretta Pettway/Artists Rights Society (ARS, NY)

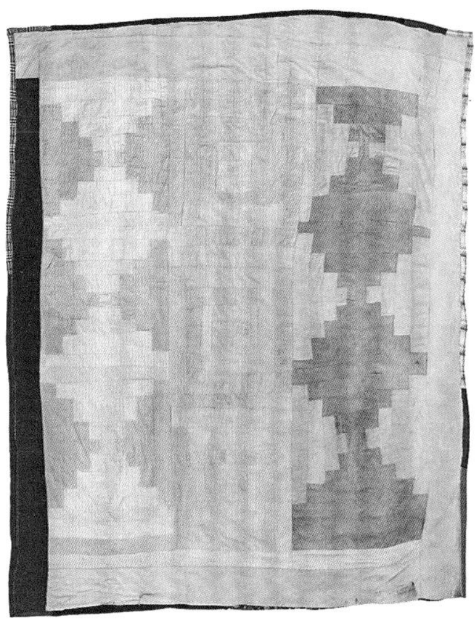

© 2019 Leola Pettway/Artists Rights Society (ARS, NY)

Jessie T. Pettway, Loretta Pettway, and Leola Pettway, African-American women of Gee's Bend, Alabama, U.S., created and constructed these beautiful quilts – breaking traditional patterns, unorthodox, fresh, amazing. Cloth quilts are *3-dimensional objects,* with textured flowing surfaces. Finally recognized as high art years later, escaping the curatorial category of 'folk art,' several quilts from the women of Gee's Bend now reside in major art museums.

IMAGE QUILTS (see ET+ Adam Schwartz, www.imagequilts.com, for examples + *IQ* code)

Image quilts gather multiple images to tell a story, make comparisons, set context. Quilt elements vary in size, have no overt frames, and abut. This quilt celebrates E. J. Marey's work on the velocity of seeing, motion pictures, data collection, visualization, and his classic *La Méthode graphique dans les sciences expérimentales* (1878).

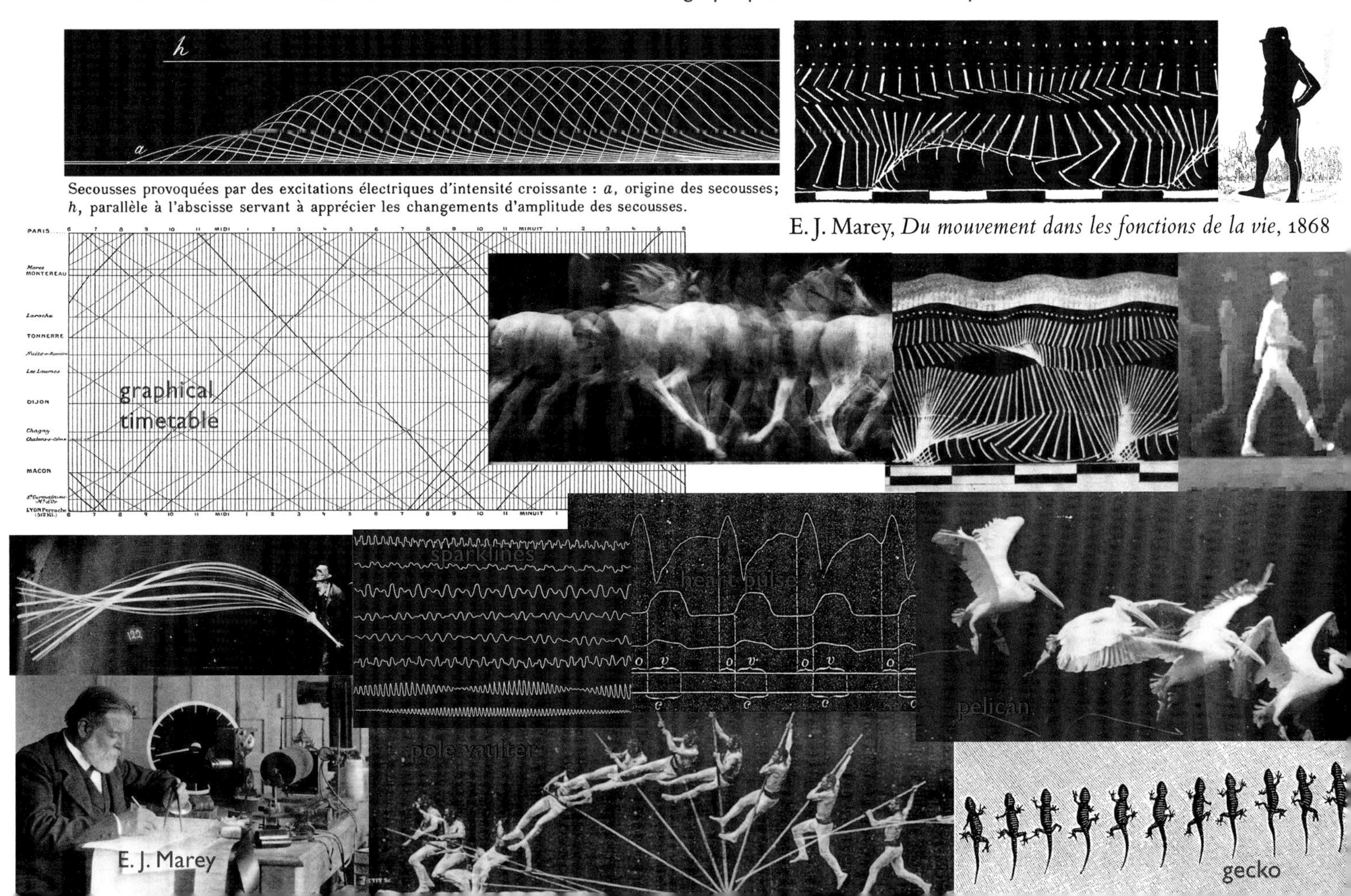

Directly adjacent images compare images – all the same size, ordered by the time these photographs were taken

Adjacent, connected, unframed image elements vary in size. This reads as Monet's farm of 15 seasonal haystacks

Frames around images enhance/isolate color and reduce the bright-white glare of display screens. Serious video and image editing are done in dark rooms with non-reflective screens to reduce screen white-noise. Over-framing of published image matrices is common, creating frame noise, reducing local space devoted to content, masking direct comparisons and flows of adjacent images. In laying out images of varying sizes, algorithms for optimizing two-dimensional packing of images are not helpful; better to use ragged margins and small black surrounds of quilts with abutting images. Over-framing of images and over-pulling of faces in plastic surgery both create a look of perpetual surprise.

47 animal sounds repeated in a sparkline matrix, ragged left/right enhance seeing:

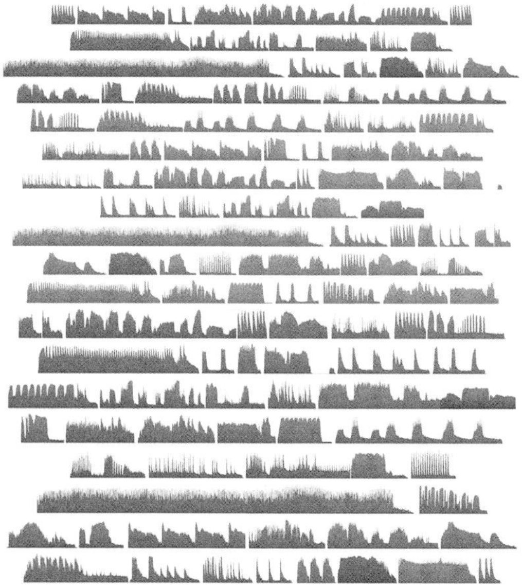

Squared-off margins not so good:

Georges Vantongerloo paintings/sculptures: ragged left/right *and* ragged top/bottom margins:

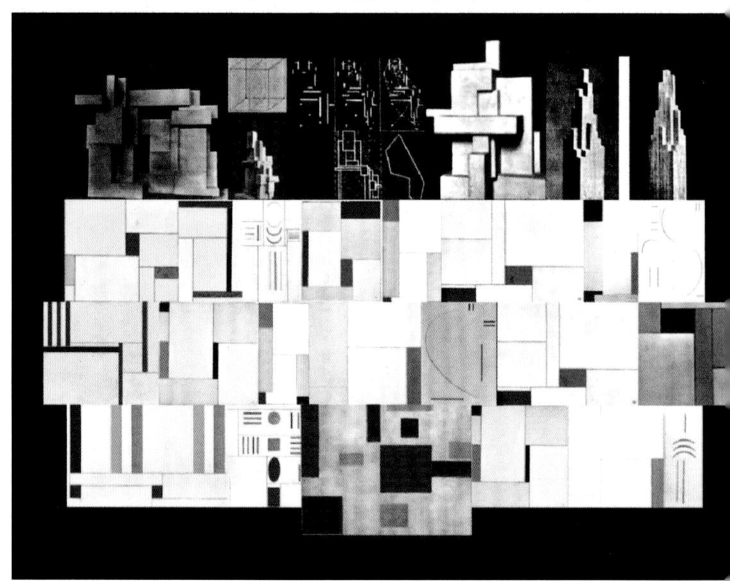

IMAGE QUILT ANALYSIS: LEARNING FROM BIG DATA SETS

Building a data set of 88 million disease instances each located in space and time 1888 to 2011, epidemiologists were able to quantify the benefits of vaccination: a total of 103 million disease cases were prevented since 1924 in the United States. This natural experiment (before vs. after) shows strong consistent effects in all U.S. states for 5 diseases, as each state serves as its own control over the years – adding up to direct visual proof. Shown here are measles vaccination data.

Willem G. van Panhuis, et al, 'Contagious Diseases in the United States from 1888 to the present,' *NEJM*, 369, 2013; Tynan DeBold and Dov Friedman, 'Battling Infectious Diseases in the 20th Century: The Impact of Vaccines,' *WSJ*, 11 February 2015

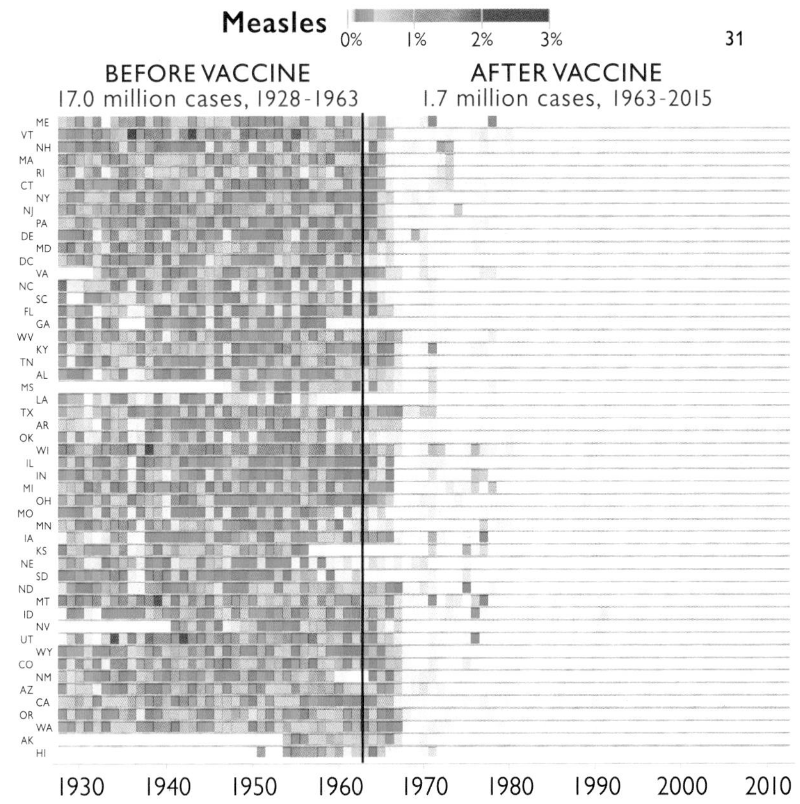

Image quilts show data within the common eyespan. Image quilts, like any evidence, must provide reasons to believe. Pay fierce attention to image data collection. Often you learn way more about a process when watching data collection at the exact moment of measurement.

This image quilt shows batch effects in sequencing genome data. Batch effects occur when unwanted variation is added to runs in data batches (genome sequencing, signal processing, scans, industrial processes). Happens all the time. Statistical adjustments for batch effects may create additional artifacts in aggregating mixed data sets. "*Batch effects for sequencing data from 1000 Genomes Project:* Each row is a HapMap sample processed in the same facility/platform: Dark blue = 3 standard deviations below average. Dark orange = 3 standard deviations above average."

Samples ordered by date (only days with extreme effects shown)

largest cluster of batch effects, days 243-251

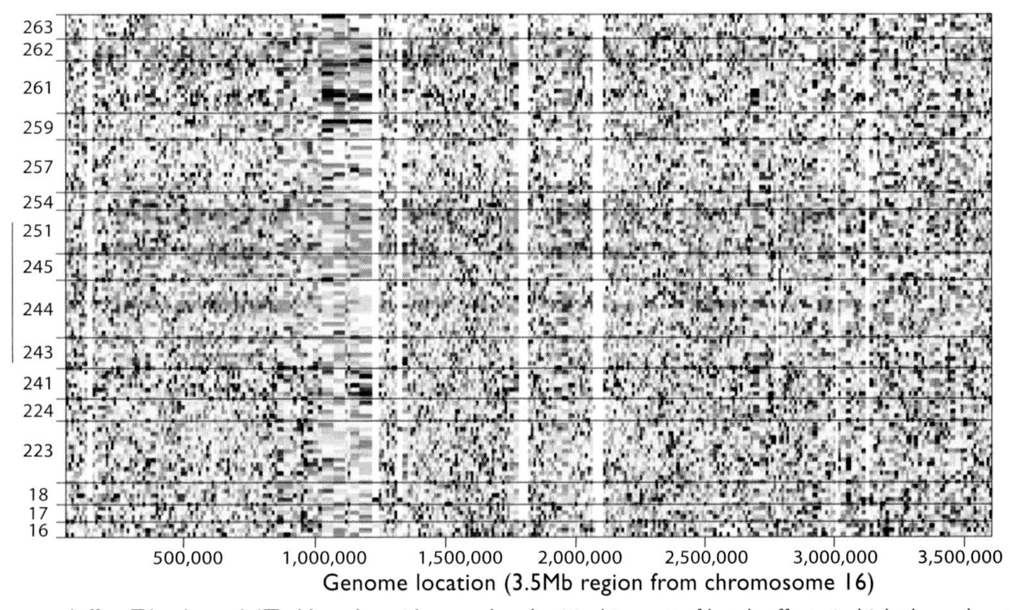

Jeffrey T. Leek, et al, "Tackling the widespread and critical impact of batch effects in high-throughput data," HHS Public Access authors manuscript, 2010, edited; *Nature Reviews Genetics* 11, 2010, 733-739.

Good cartography demonstrates *practical design strategies* for showing data: high resolution, color scales, multiple layers, contour lines, words, numbers located anywhere everywhere on a substrate grid of meaning. But the point of view of a map is from above, often far above. This may not help people moving around on the ground.

WHEN THE MAP MODEL BREAKS DOWN: INSTEAD OF A UNIFORM FLATLAND GRID, SOMETIMES A BETTER MODEL IS MULTIPLE LOCAL VIEWS ORDERED IN 2-SPACE

Maps and aerial photographs directly translate real-space into map-space units, with no local grids. In contrast, this flamboyant confection by Google Maps locates street views on the substrate of an aerial photograph, but then shows 19 traffic signs as seen in position by San Francisco drivers. Cézanne imaged 4 points of view, Google 19, Hockney 750, the Madagascar ground gecko embryo 12,000. Living, moving eyes routinely register 1000s of viewpoints per hour anywhere everywhere. That is what eyes do in 3-space and time – scanning broadly, then focusing on locally – from all sorts of points of view. Can we see in the real world as closely as when we examine a data graphic, read a poem, watch a video?

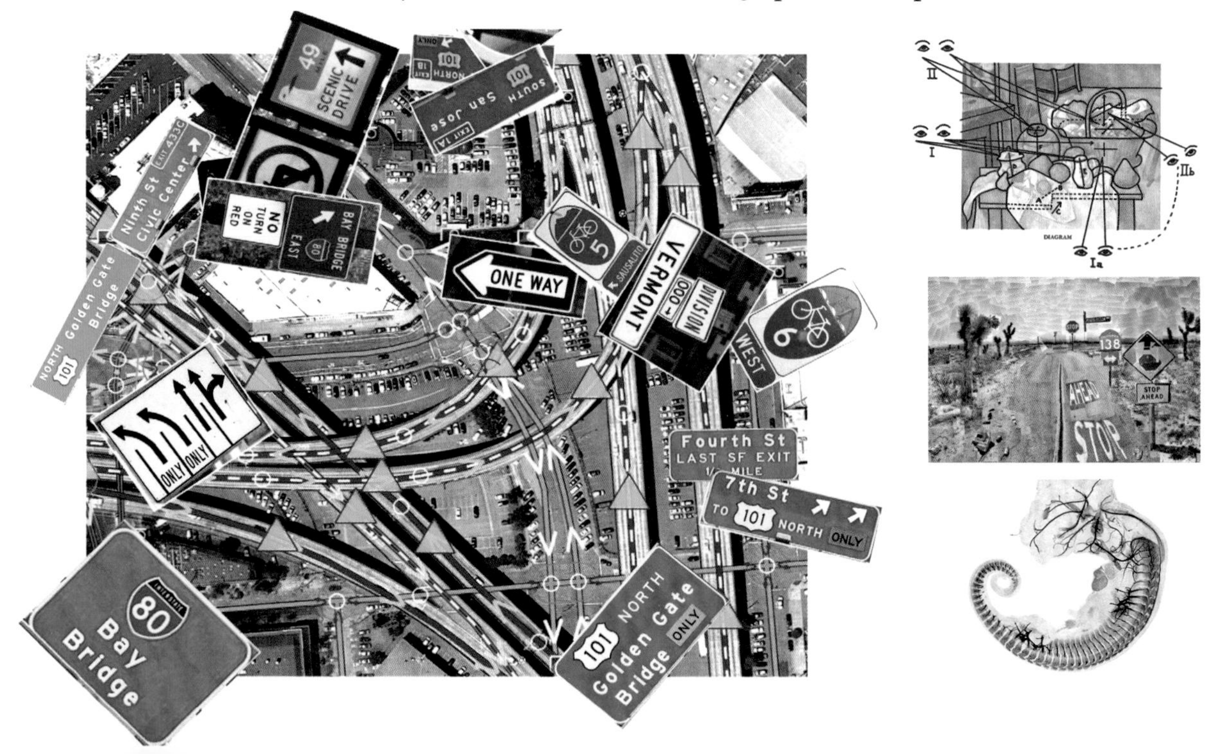

LOCAL VIEWS ORDERED IN SPACE: VISITORS GUIDE TO ET OUTDOOR ARTWORKS

A visual guide for my landscape sculptures started out, wrongly, with an aerial photo. Land is vast compared to my artworks, and aerial views provide only addresses of tiny images seen from above. Instead, this quilt shows local points of view, ordered on a north/south by west/east substrate (scaled by objects of known size: sheep, people, Ace the dog, 0.7 meters). Walking around, our visitors navigate by what they actually see – artworks, hills, distances, land, ponds. Outdoor art creates endless fresh views, all for free, as light, weather, viewer locations change. I created these artworks and Hogpen Hill Farms artpark 2008-2021 and continuing.

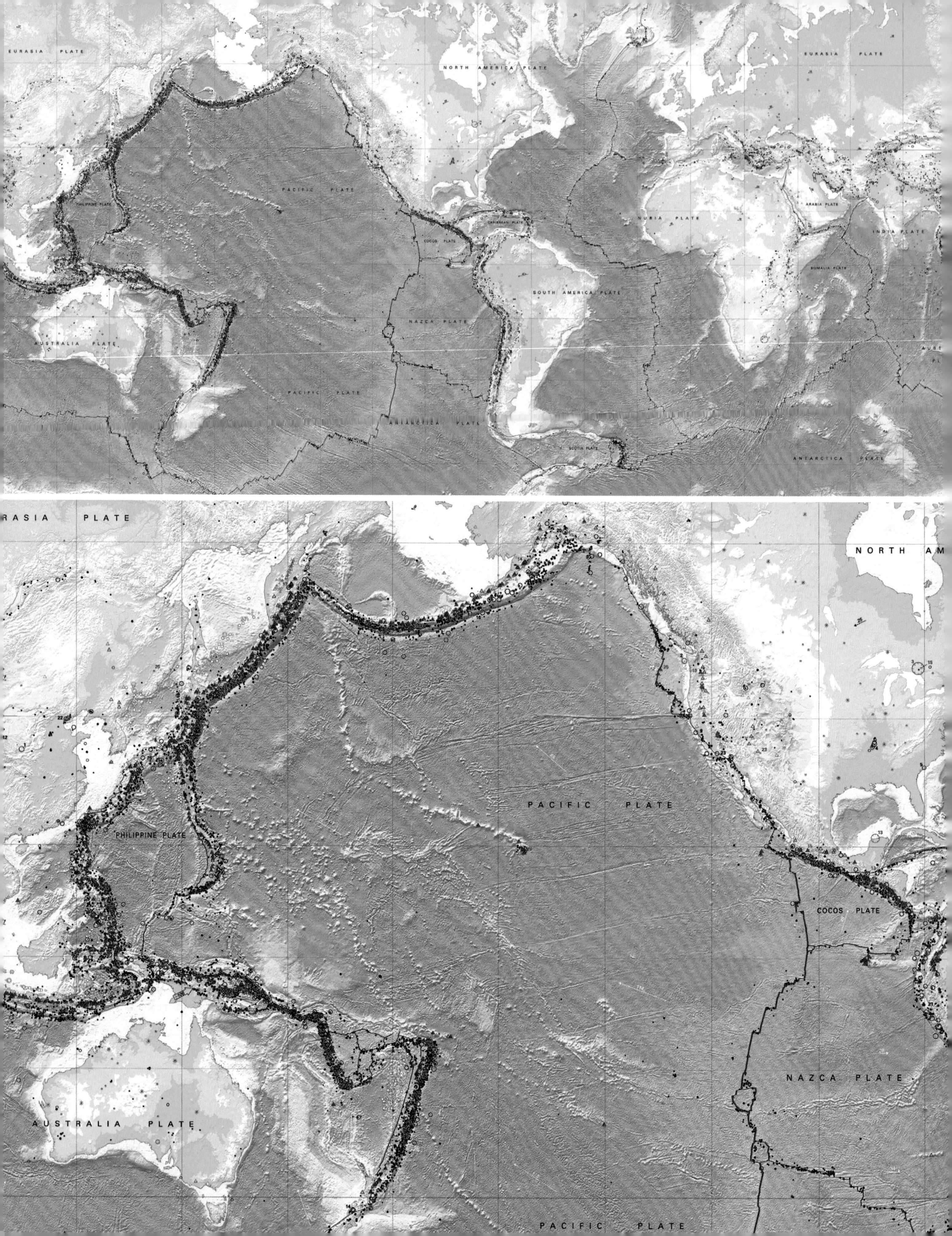
EURASIA PLATE
NORTH AMERICA PLATE
EURASIA PLATE
PACIFIC PLATE
PHILIPPINE PLATE
CARIBBEAN PLATE
COCOS PLATE
NUBIA PLATE
ARABIA PLATE
INDIA PLATE
SOMALIA PLATE
SOUTH AMERICA PLATE
NAZCA PLATE
AUSTRALIA PLATE
PACIFIC PLATE
SCOTIA PLATE
ANTARCTICA PLATE
RASIA PLATE
NORTH AM
PACIFIC PLATE
PHILIPPINE PLATE
COCOS PLATE
NAZCA PLATE
AUSTRALIA PLATE
PACIFIC PLATE

ON THE BORDERLINE AND BORDERSPACE: MARGINS, EDGES, FRAMES, COLLISIONS

Eye-brain systems routinely detect and reason about
edges margins perimeters contours outlines fences boundaries
transitions changes differences convergences divergences collisions

At the borderline and the borderspace between
meaning and space presence and absence figure and ground this and that us and them
where verbs/actions/interactions/conflicts all thrive
and eyes see/think/design/produce.

Adventures happen at borderlines in flatland and space-time worlds. Eye-brain systems, painters, and software for anti-aliasing/pattern-recognition/image-sharpening/compression all give special attention to edges or apparent edges. Real-world 3D borderspaces – where tectonic plates collide – are depicted in the classic maps at left. Tectonic plates move 20-100 mm/year,

0 mm 20 mm 100 mm

and converge/fracture/overlap, as earthquakes and volcanoes erupt along borderspaces.

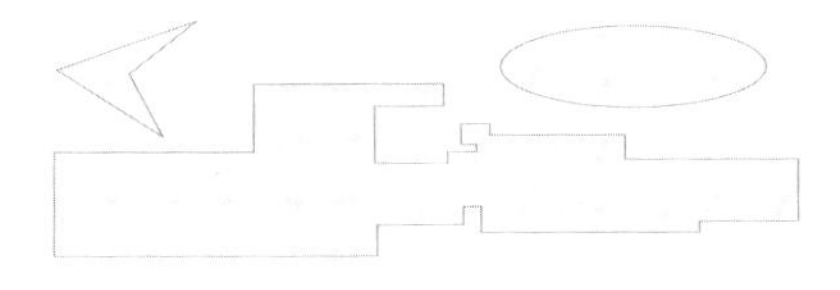

These thin framing lines produce optical interactions in the minds of viewers, for the apparent differences between gray fields are illusory. Creating effective optical consequences, thin outlines accent edges, activate contrasts/discontinuities between figure/ground, and deftly shift the perceived color values of outlined fields.

Line-and-tone maps use finely engraved outlines to frame hand-colored transparent tones, a beautiful effect. These lines/frames are content-responsive: if buildings change, lines change. Slightly thicker lines accent 2 sides of a building and create soft 3D optical effects, now and then causing the tones to read equivocally, as if indicating basements:

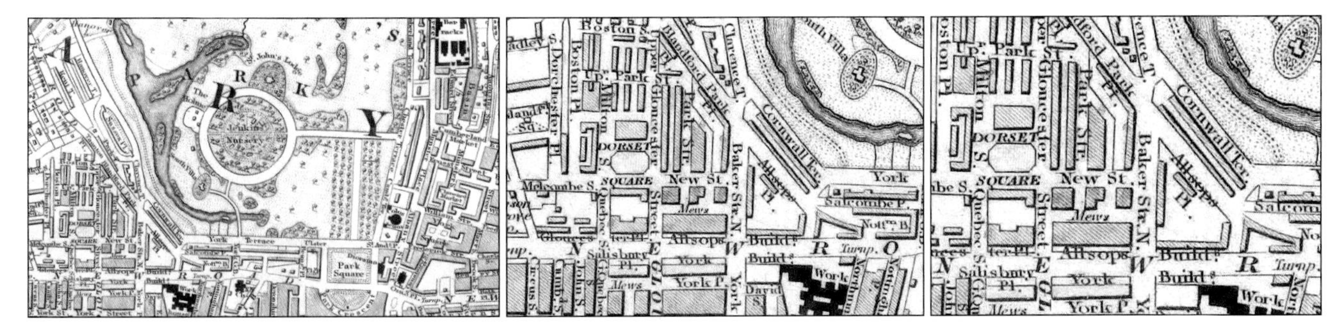

All day long mapmakers illustrators painters typographers architects sculptors tectonic plates construct and contrive interactions at borders that shape space, enhance meaning, create real/apparent surfaces and volumes, strengthen internal unity of fields, separate inside from outside.

ET, *Larkin's Twig* 2004
steel height 32 feet or 10 meters
footprint 59 x 66 x 70 feet or 18 x 20 x 21 meters

3-DIMENSIONAL TRAJECTORY LINES OF STEEL

Torquing through 3-space, the steel generates subtle transparent optical surfaces. Above, my *Larkin's Twig* (10m height, constructed from .1% steel, 99.9% air by volume). Gently torquing air surfaces appear perfectly fitted in the spaces formed by the steel legs – line-and-tone effects in 3-space. In visits to our sculpture fields, most observers can be talked into seeing these transparent nearly invisible sail-like air surfaces. Below, the effect enhanced:

AIR AND SPACE ARE MATERIALS, THE NOTHING THAT IS REALLY SOMETHING.

VIEWERS MOVING THROUGH AIRSPACE, PERCEIVED EDGES ARE SPACES, NOT LINES,

JUST AS THE HORIZON IS A SPACE, NOT A LINE

RICHARD SERRA   JUNCTION 2011   4 × 22.9 × 15.2 METERS   GAGOSIAN GALLERY, NEW YORK

I consider space to be a material.

The articulation of space has come to take precedence over other concerns.

I attempt to use sculptural form to make space distinct.

RICHARD SERRA

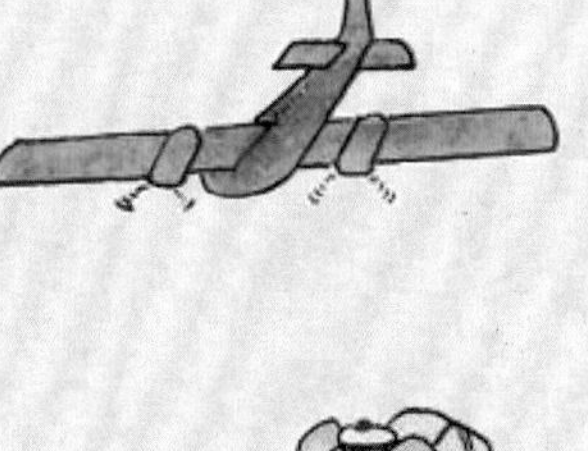

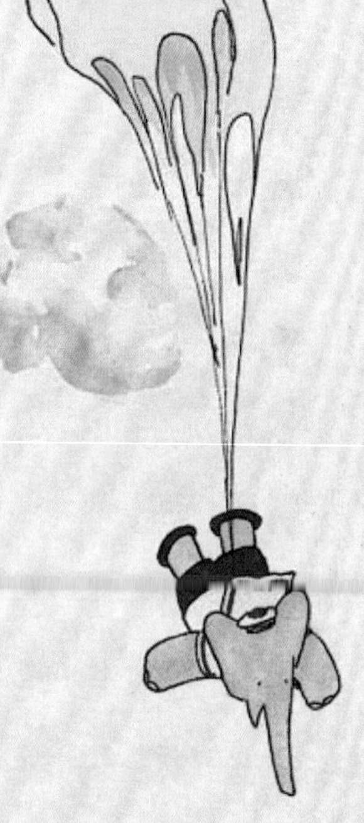

MOVING IN SPACE, EVOLVING IN TIME, SEQUENTIAL

STOP-ACTION IMAGES REVEAL LOCAL EVENTS IN A SCENIC CONTEXT

To explore/understand/explain dynamic flowing information, stop-action images adjacent in space are helpful – and often better than continuous video, where the quick pace and high autocorrelation blurs analytical thinking. These well chosen stop-action images illustrate swinging-pendulum and spinning-rotation effects during a parachute descent of Arthur, Babar's cousin. Unusual information architectures can be found by looking around in different places, here in a printer's proof for a children's book by Laurent de Brunhoff, *Babar et ce coquin d'Arthur,* Paris, 1946.

When images are ordered historically, as in these Monet haystack paintings, time and space become synchronized and coherent. This stop-action timeline below shows 9 months of Monet painting light outdoors.

ROCKET SCIENCE #3

AIRSTREAM INTERPLANETARY EXPLORER

Edward Tufte

length 84 feet or 26 meters

height 31 feet or 9 meters

Since Feynman diagrams describe the universal operations of Nature's laws, they can communicate throughout the universe. Both sides of the *Airstream Interplanetary Explorer* show Feynman diagrams that might communicate with intelligent life anywhere. Better the cosmopolitan verbs of Nature's laws on spacecraft than the local proper nouns of national flags, earthly Gods and Goddesses, and government agency logos. For interplanetary exploration, better to send smart machines and their emblematic Feynman diagrams than human beings and their lawn chairs, toilets, teddie bears.

For the cosmological entertainment of intelligent beings wherever whenever, a prankish illusory violation of Nature's laws makes a rare joke that might travel well – as in the Pioneer Space Plaque redesign at right and far right.

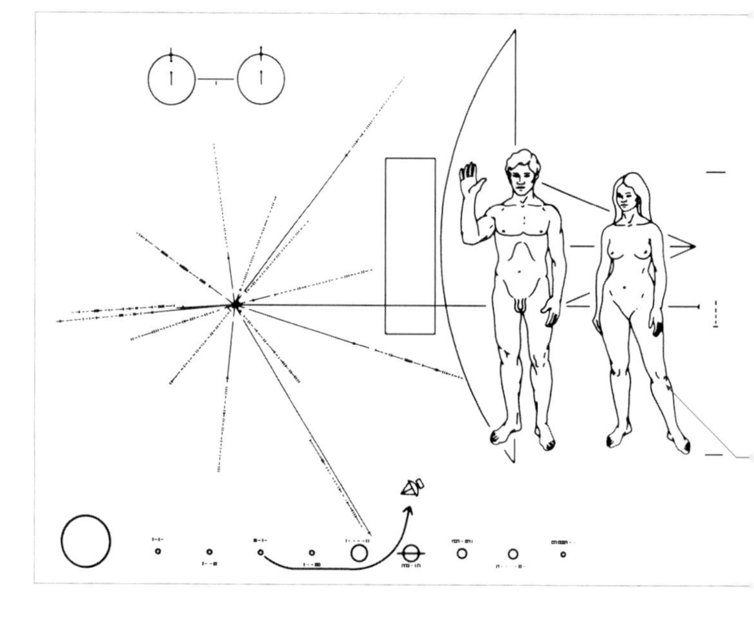

ROCKET SCIENCE, REMODELED

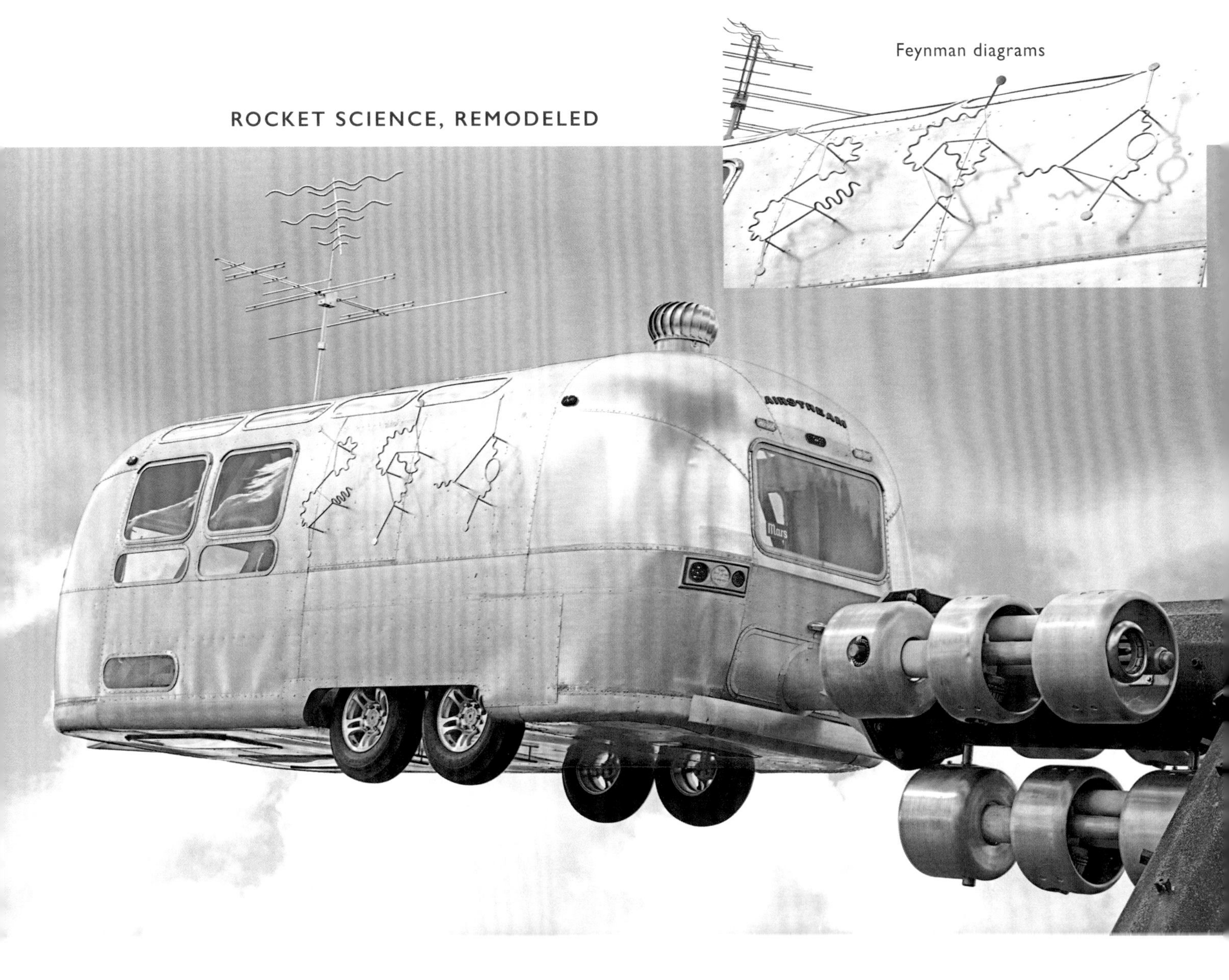

THE PIONEER SPACE PLAQUE: A COSMIC PRANK

Edward Tufte   digital print, animation electronics   6.9 × 2.2 × .5 feet or 2.1 × .7 × .2 meters

Magic, the production of entertaining illusions, has an appeal quite independent of the local specifics of language, history, or culture. In vanishing objects or levitating assistants, conjurers amaze, delight, and even shock their audiences by the apparent violation of the universal laws of nature and our daily experience of those laws.

Since the principles of physics hold everywhere in the entire universe, magic is conceivably a cosmological entertainment, with the wonder induced by theatrical illusions available to and appreciated by all, regardless of planetary system. Accordingly the original plaque placed aboard the Pioneer spacecraft for extraterrestrial scrutiny billions of years from now might have escaped from its conspicuously anthropocentric gestures by showing instead the universally familiar Amazing Levitation Trick.

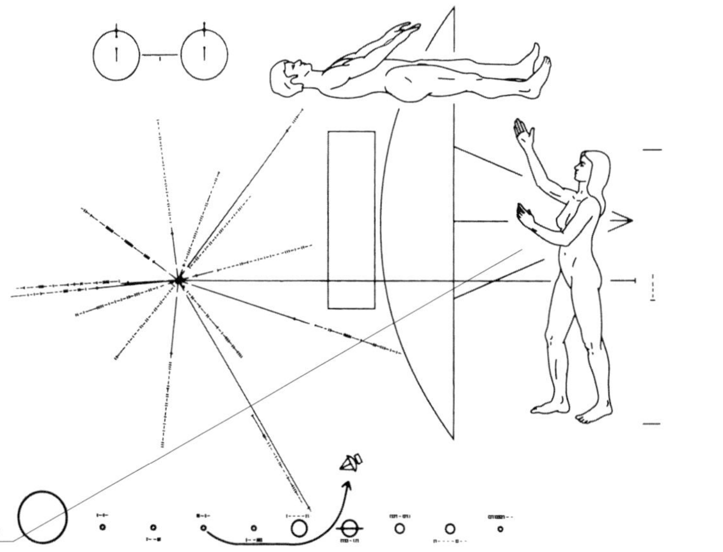

THE GREAT REMODELING OF ART:

ON-THE-GRID REPRESENTATIVE ART REPLACED BY OFF-THE-GRID ABSTRACT ART

HOW TO VIEW HIGH (ABSTRACT) ART

". . . People who try to explain pictures are usually barking up the wrong tree." —Picasso

When you "commune with nature" you enjoy, at first hand, beautiful natural sounds that are not "man-made."

Some "man-made" sounds (opera, program music, etc.) imitate—or describe, natural scenes, actions, or something.

A lot of "man-made" sounds don't try to imitate—they are to be enjoyed for their own rhythms and structures.

When you "commune with nature" you also enjoy, at first hand, beautiful natural sights that are not "man-made." ("Only God can make a tree.")

Ad Reinhardt, "How to view high (abstract) art," published in *P.M.* daily newspaper, February 24, 1946

NATURE, SOUNDS, ABSTRACT ART RESIDE ON LOCAL GRIDS

SLIP THE CANDID CAMERA TO HIM, JOHN BOY
NEW JERSEY
TITLES
Some "man-made" sights (picture-paintings, illustrations, etc.) imitate or try to reproduce or remind you of natural scenes and associations (second-hand).
HA HA WHAT DOES THIS REPRESENT?
WHAT DO YOU REPRESENT?
HOW TO PAINT
LIST OF EXHIBITIONS
MODERN ART
ART HISTORY
REINHARDT
Some "man-made" sights (abstract, non-objective paintings) try to recreate new relationships of lines, colors and spaces and are to be enjoyed as such.

Ad Reinhardt, "How to view high (abstract) art," published in *P.M.* daily newspaper, February 24, 1946

Whatever-it-takes is the design method for Ad Reinhardt's *How to View High (Abstract) Art.* This 6-panel comic shows annotated word-thoughts and image-thoughts. In collages of clip art, pieces of type, new and old drawings, Reinhardt compares representative art (perspective grid) vs. abstract art (off the grid). And then forcefully points out that in music and Nature, we hear/see/enjoy representations *and* abstractions. Thus for abstract art as well, which creates "new relationships of lines/colors/spaces are to be enjoyed as such."

In *The Silence of Animals,* John Gray deploys the same concepts as Ad Reinhardt – words/space/silence/sounds/nature/meaning/images/perception. But instead of advancing a Defense of High Abstract Art (off the grid, often beyond words), John Gray arrives at Intense Contemplation (off the grid, and way beyond words):

Philosophers will say that humans can never be silent because the mind is made of words.
For these half-witted logicians, silence is no more than a word.
To overcome language by means of language is obviously impossible.
Turning within, you will find only words and images that are parts of yourself.
But if you turn outside yourself –
to the birds and animals and the quickly changing places where they live –
you may hear something beyond words. Even humans can find silence,
if they can bring themselves to forget the silence they are looking for.

JOHN GRAY

EMPIRICAL REMODELING IS OPEN-MINDED AND RIGOROUS

Remodeling often begins by finding metaphors that transform prevailing models, such as replacing hierarchies with webs, reforming linear sentences with the content-driven linebreaks of poetry and coding, and making interfaces and data displays face up to maps. Reasoning about metaphors requires direct specific comparisons, not vague handwaving. Remodeling requires an open searching mind and at the same time rigorous standards of evidence and judgment. It is even more difficult to hold in mind simultaneously and without premature judgment your own ideas and opposing ideas. Few are able to do so.

TO HAVE AN
OPEN MIND
BUT NOT AN
EMPTY HEAD
– TUFTE

AUSTIN KLEON

SEEING WITH FRESH EYES: A SENSE OF THE RELEVANT

Thinking/seeing/analytical/fresh eyes have a sense of what is relevant, finding what is important in a mass of data and knowledge, seeing something worthwhile, where to gain leverage. A sense of the relevant is the ability to identify and detect those things that have consequences beyond themselves.

The common guide to relevance is a profession or discipline. There are many good reasons for local expertise. That's why it is called a discipline. But from a reforming and creative point of view, my point of view, the world is so much more interesting/amazing/consequential than any one discipline or profession.

Creativity is connecting things. Connections become enshrined, narrowed, exhausted, prohibited within a discipline or specialty. Fresh eyes approach a new field with deep curiosity and focused looting – to quarry, make new connections, do whatever it takes, seek the useful and relevant, learn afresh not merely confirm prior views. When you meet with experts in another field, shut up and listen. People love to talk about their stuff, and your job is to provide occasional prompts to guide them toward your interests, to take notes, now and then testing their explanatory depth, seeking to understand their cognitive style, asking 'Why is that?' 'How do you know that?' Be skeptical, learn what you can about a field but don't believe it. Outsiders need not accept the assumptions and doctrines of disciplines.

SEEING WITH FRESH EYES: IDENTIFY AND KNOW EXCELLENCE

Learn what excellence is, how to identify it. Learn about cognitive styles and strategies of excellent work, what credible reasoning looks like. This is not a big reading assignment – excellence is scarce, lognormal, long-tailed. Acting on this knowledge is liberating, freeing oneself from vast piles of triviality, knock-offs, petty connoisseurship, over-publishing, and the short-sighted, trendy, greedy. Excellence is long-term knowledge, even forever knowledge.

Excellence, like good taste, is perhaps a universal quality. Analytical thinking is about the relationship between evidence and conclusions, and is fundamental to all empirical work, regardless of field, discipline, specialty. Thus it is possible at times to assess credibility of nonfiction work without being a content expert. Thinking eyes may well have an eye for excellence, regardless of field or discipline.

And what is
if you know once you
think you know?

MAIRA KALMAN

What do we know that is not true?

CHRIS QUIGG

SELF-POISONING: ARE WE ALL VIGILANTE, AMATEUR COPY EDITORS?

All writers/readers collect rules about grammar/typography/punctuation/word-usage/spelling. Many regard their rules as absolute and righteous. All different, personal style-sheets create chronic background nuisance noise distractions in writing/reading/editing. Conventions are necessities, but stay focused on the content, and stop arguing with the author's private mash-up style sheet vs. your private mash-up style sheet. Why use your private mashup stylesheet as a proxy for assessing content credibility? Conventions and guidelines increase outsider entry costs, preserve insider status quo, cause long-run inequities. Infinitive splitters rarely get into Harvard or Oxford.

CONTENT-RESPONSIVE TYPOGRAPHY PREVENTS BILLIONS (10^9) OF IMPEDIMENTS TO READING DARWIN'S *ORIGIN OF SPECIES*

The title page of Charles Darwin's *The Origin of Species* uses central-axis typography with natural content linebreaks. The analytical Table of Contents then announces 14 chapters with a total of 133 topics. Each chapter also begins with a topic list, a repeat from the Table of Contents. These topic lists provide 1,962 words summarizing *Origin*. The typography creates many impediments to reading a book that many scientists regard as the most important scientific discovery ever. These topic lists suffer from content-hostile linebreaks, grid-driven squaring off the right margin and mushing the section titles together. Here, the topic list for the famous concluding chapter where Darwin begins by considering difficulties in the theory of natural selection:

ON

THE ORIGIN OF SPECIES

BY MEANS OF NATURAL SELECTION,

OR THE

PRESERVATION OF FAVOURED RACES IN THE STRUGGLE
FOR LIFE.

BY CHARLES DARWIN, M.A.,

FELLOW OF THE ROYAL, GEOLOGICAL, LINNÆAN, ETC., SOCIETIES;
AUTHOR OF 'JOURNAL OF RESEARCHES DURING H. M. S. BEAGLE'S VOYAGE
ROUND THE WORLD.'

LONDON:
JOHN MURRAY, ALBEMARLE STREET.
1859.

CONVENTIONAL TEXT TYPOGRAPHY,
BASED ON THE NEEDS OF LEAD TYPE,
SUFFERS FROM LEAD POISONING

Recapitulation of the difficulties on the theory of Natural Selection — Recapitulation of the general and special circumstances in its favour — Causes of the general belief in the *immutability of* species — How far the theory of natural selection may be extended — Effects of its adoption on the study of Natural history — Concluding remarks

CONTENT-RESPONSIVE TYPOGRAPHY
CREATES CONTENT LINE-BREAKS,
IS MORE READABLE AND NATURAL

Recapitulation of the difficulties on the theory of Natural Selection
Recapitulation of the general and special circumstances in its favour
Causes of the general belief in the immutability of species
How far the theory of natural selection may be extended
Effects of its adoption on the study of Natural history
Concluding remarks

Content-responsive typography clarifies topic paragraphs for both *Origin's* table of contents and chapter headings. These changes add 2 or 3 pages to *Origin's* 500 pages. Eliminating typographic impediments does add up: 133 topics, each appearing twice, perhaps 3 million serious readers and millions of readers who skimmed over *Origin of Species* since 1859 – a grand total of ~2,000,000,000 to ~5,000,000,000 typographic impediments for *Origin's* readers. Content-responsive typography can assist both readers and skimmers.

CONTENT-POSSESSED TYPOGRAPHY: APOLLINAIRE'S CALLIGRAMME *IL PLEUT* (RAIN)

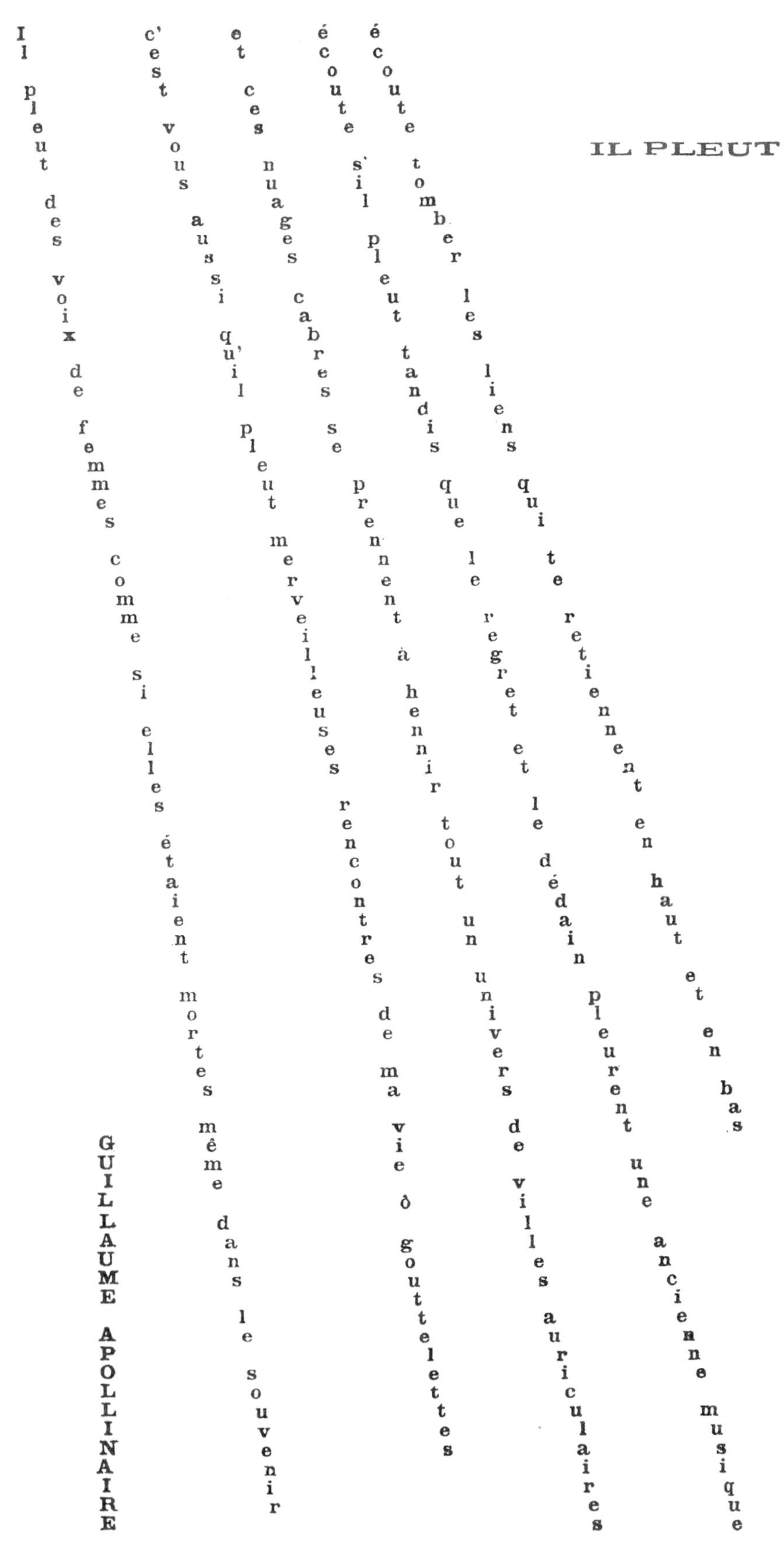

CONVENTIONAL TYPOGRAPHY
Content-indifferent word-spaces and line-breaks are manipulated to square off the right margin and save space. Syntactical structure is lost, creating a word salad. And why only words, why segregate information by its mode of production? Why not also include small pictures perhaps taken by the author of stacks/shelves of the actual books, their spines showing?

There were books of all kinds. Volumes of verse, novels, philosophical works, critical studies (they say books about books are profitless, but they certainly make very pleasant reading), biographies, history; there were books to read when you were ill and books to read when your brain, all alert, craved for something to grapple with; there were books that you had always wanted to read, but in the hurry of life at home had never found time to; there were books to read at sea when you were meandering through narrow waters on a tramp steamer, and there were books for bad weather when your whole cabin creaked and you had to wedge yourself in your bunk in order not to fall out; there were books chosen solely for their length, which you took with you when on some expedition you had to travel light, and there were the books you could read when you could read nothing else.

CONTENT-RESPONSIVE TYPOGRAPHY
Line-spacing and linebreaks clarify meaning, quietly and effectively. Commas, semi-colons, and full stops in the original text signal a linebreak for the content-responsive text.

More imaginative linebreaks result from closer readings of the text, discovery via trial and error – and thinking like a poet about meaning in linebreaks, punctuation, enjambment.

There were books of all kinds

Volumes of verse, novels, philosophical works, critical studies
(they say books about books are profitless, but they certainly make very pleasant reading)
biographies, history

There were books to read when you were ill
and books to read when your brain, all alert,
craved for something to grapple with

There were books that you had always wanted to read
but in the hurry of life at home had never found time to

There were books to read at sea
when you were meandering through narrow waters on a tramp steamer
there were books for bad weather when your whole cabin creaked
and you had to wedge yourself in your bunk in order not to fall out

There were books chosen solely for their length
which you took with you when on some expedition you had to travel light
and there were the books you could read when you could read nothing else

CONVENTIONAL TYPOGRAPHY
It is difficult to identify and separate out the 6 list elements. The 5 semi-colons must work hard to signal each one of the 6 elements.

Why Most Published Research Findings Are False: when the studies conducted in a field are smaller; when effect sizes are smaller; when there is a greater number and lesser preselection of tested relationships; where there is greater flexibility in designs, definitions, outcomes, and analytical modes; when there is greater financial and other interest and prejudice; when more teams are involved in a field in chase of statistical significance. John Ioannidis

CONTENT-RESPONSIVE TYPOGRAPHY
Separate linebreaks identify each list-element. Readers compare and remember list-elements by scanning up/down, impossible in ordinary text. Spaces clarify meaning quietly and effectively.

Why Most Published Research Findings Are False:
when the studies conducted in a field are smaller,
when effect sizes are smaller,
when there is a greater number and lesser preselection of tested relationships,
where there is greater flexibility in designs, definitions, outcomes, analytical modes,
when there is greater financial and other interest and prejudice,
when more teams are involved in a field in chase of statistical significance.

CONVENTIONAL TEXT
This paragraph describes fresh, intriguing, complex ideas. But the 6 sentences do run together, masking distinctions between vacuum and void.

Vacuum, in modern physics, is what you get when you remove everything you can, whether practically or in principle. Alternatively, vacuum is the state of minimum energy. Intergalactic space is a good approximation to a vacuum. Void, on the other hand, is a theoretical idealization. It means nothingness: space without independent properties, whose only role is to keep everything from happening in the same place. Void gives subatomic particles addresses, nothing more.

CONTENT-RESPONSIVE
The first 3 sentences describe *Vacuum,* the second 3 describe *Void.* A line space separates *Vacuum* vs. *Void* sentences, clarifying this distinction. All 6 line-breaks are content-driven, enhancing meaning. Read again to see the linebreaks at work.

Vacuum, in modern physics, is what you get when you remove everything you can,
whether practically or in principle. Alternatively, vacuum is the state of minimum energy.
Intergalactic space is a good approximation to a vacuum.

Void, on the other hand, is a theoretical idealization. It means nothingness:
space without independent properties, whose only role is to keep everything from
happening in the same place. Void gives subatomic particles addresses, nothing more.

This sparkling conclusion causes thoughts few readers have ever thought: void keeps everything from happening in the same place, gives subatomic particles addresses. *Italic letters – extending letters diagonally, compressing horizontally – enhance flowing words.* Ceremonial central-axis typography reinforces meaning, a paragraph of Annunciations and Revelations.

Central-axis (flush center) clearly signals each line, and readers/speakers don't have to search on the left margin, sometimes accidentally skipping down a line. In central-axis, each line is activated at both left and right margins – unlike squared-off conventional text. Readers/speakers are aware of line-length at both beginning/end. That knowledge accents meaning and when reading aloud.

50 YEARS AGO, HYPERLINKS MADE ELECTRONIC TEXT MORE CONTENT-RESPONSIVE

Hypertext is simply electronic text containing direct links to other texts.
By clicking on links in hypertext documents, readers can immediately link to other content anywhere.

Hypermedia is not restricted to text, and can include links to graphics, video, sound, etc.

Hypertext Transfer Protocol (HTTP) underlies the WorldWideWeb protocol for distributed/collaborative/hypermedia information systems. HTTP is the foundation of data communication for the WorldWideWeb, where hypertext documents include hyperlinks to other resources that users/readers/thinkers can directly access, by clicking or tapping a web browser screen.

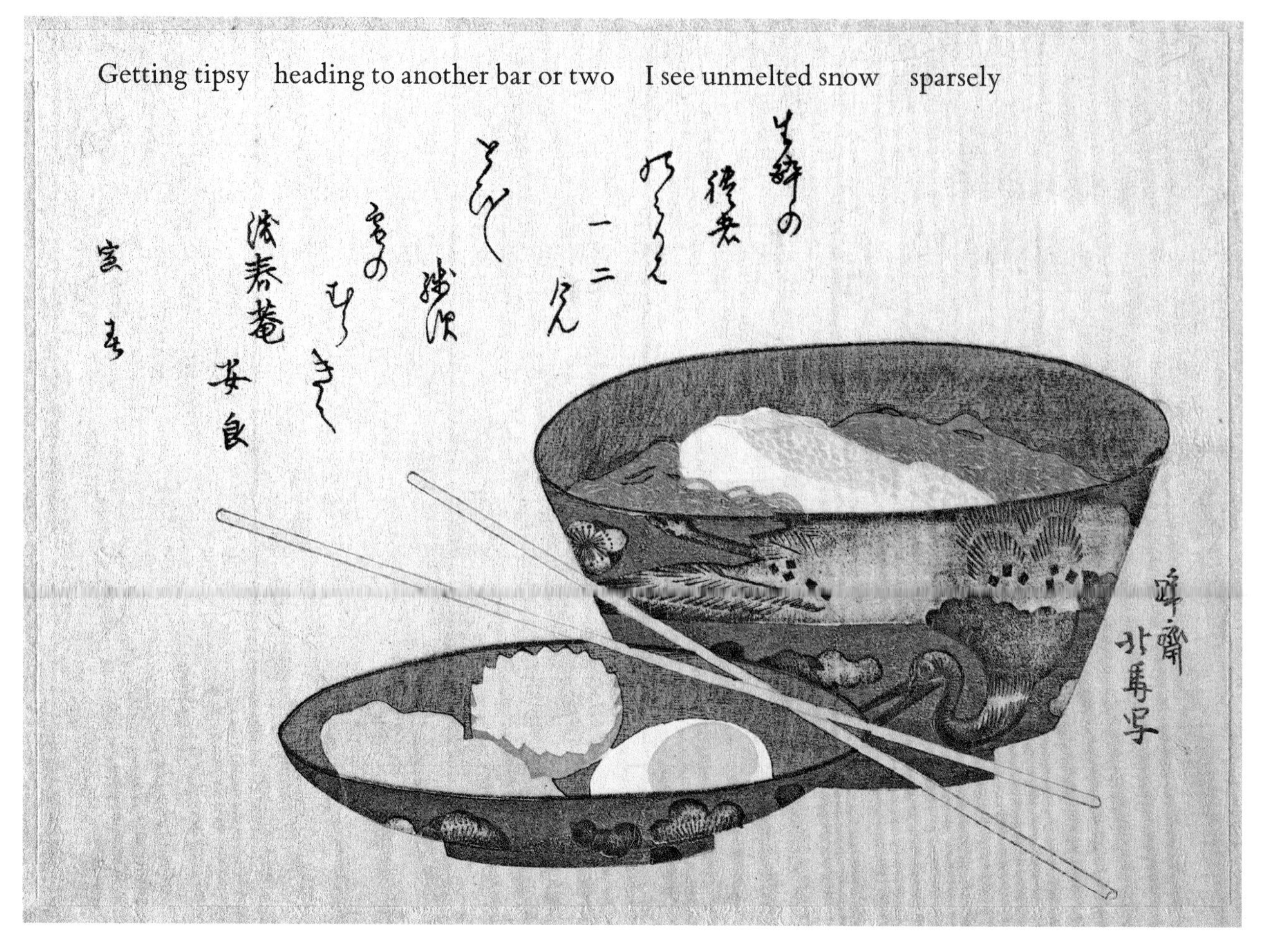

Getting tipsy heading to another bar or two I see unmelted snow sparsely

MEANING AND SPACE: SENTENCES IN 2-SPACE

Teisai Hokuba, *Bowl of New Year Food* (woodblock print, 1808): tipsy characters wander through perspective space above the bowls and chopsticks, which create receding planes that accent the drifting calligraphy. These calligraphic figures enliven both content and space.

SEQUENTIAL STOP-ACTION, WORDS, ILLUSTRATIONS IN THIS LUSCIOUS DATA PARAGRAPH

EGYPTIAN VULTURE

The outline is most easily controlled if one begins with the head, then draws the front of the bird, continuing with the rearward leg. If this much is correctly executed, it is relatively easy to add the rest. Note the sharp angle at the back of the head, which is explained by the fact that the feathers in that region tend to stand out, particularly if fanned by a breeze.

This multimodal confection combines 4 data sources: a vulture line-and-tone illustration, 2 line drawings comparing the Egyptian and modern vulture, and elaborate instructions at point of need 4000 years ago for drawing a vulture hieroglyph (3 sentences, 6 stop-action sketches demonstrating correct hieroglyphic stroke-order). A lovely concluding sentence describes vulture headfeathers – 'particularly if fanned by a breeze.'

IMPLEMENTING CONTENT-RESPONSIVE TYPOGRAPHY: CAPTIONS AND ANNOTATIONS

SUTURING CONNECTS 2 ARTERIES

LETTER-CODE CAPTIONS: INCONVENIENT TO CONTENT AND READERS

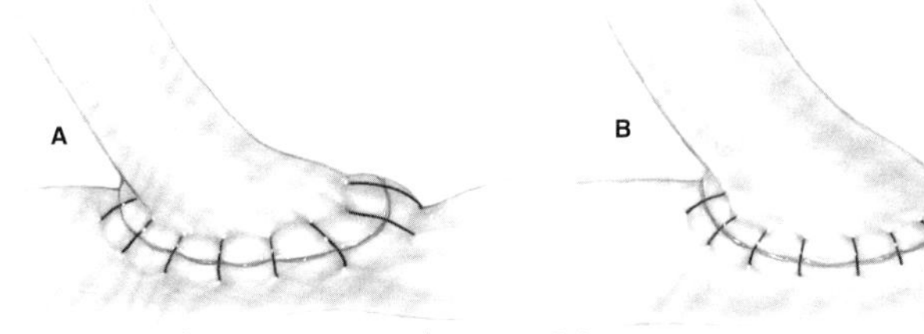

A: Dimpling and narrowing at the toe of the anastomosis.
B: Small, close-together suturing at the toe prevents narrowing of the anastomosis.

CONTENT-HINTED TYPOGRAPHY: DIRECT LABELS, JUST LIKE A MAP

⊘ Dimpling and narrowing at anastomosis toe

☆ Small, close-together suturing at anastomosis toe prevents dimpling and narrowing

IMAGE CAPTIONING: INTENSE CONTENT-RESPONSIVE DOUBLE ANNOTATIONS

The Association for Vascular Access, a U.S. medical trade group, annotated (black type) a photograph of tennis star Andy Murray's post-operative intravenous lines in a London hospital. The AVA suggested IV lines should be changed: **two PIVs...in the same arm and possibly in the same vein?**

Are these claims correct, or just a speciality trade association pitching their guidelines (see chapter 4)? Changing IV lines causes infections; current good practice is to change every 3 or 4 days, rather than risking daily needles.

British doctor Peter Lax then brightly replied (red type) to AVA annotations. For example: **Not when in left lateral position for hip surgery** **blood pressure cuff should be on the IV-free arm**

Not when in left lateral position for hip surgery
blood pressure cuff should be on the IV-free arm

Arm hair should not be clipped as it increases infection rate...
arm hair should be clipped

Dressings themselves are occlusive, the tape is to stop lines being pulled out
dressings are not occlusive and should be changed

blood visible in IV tubing
Common immediately post op and not concerning

stopcocks should be removed immediately following surgery. They do not disinfect and are known to increase infection risk
Common immediately post op and not concerning

Needed for operational positioning and not a problem
two PIVs...in the same arm and possibly in the same vein?

Patient is a right-handed professional tennis player; consider cannulating his left arm instead
Not when in left lateral position for hip surgery

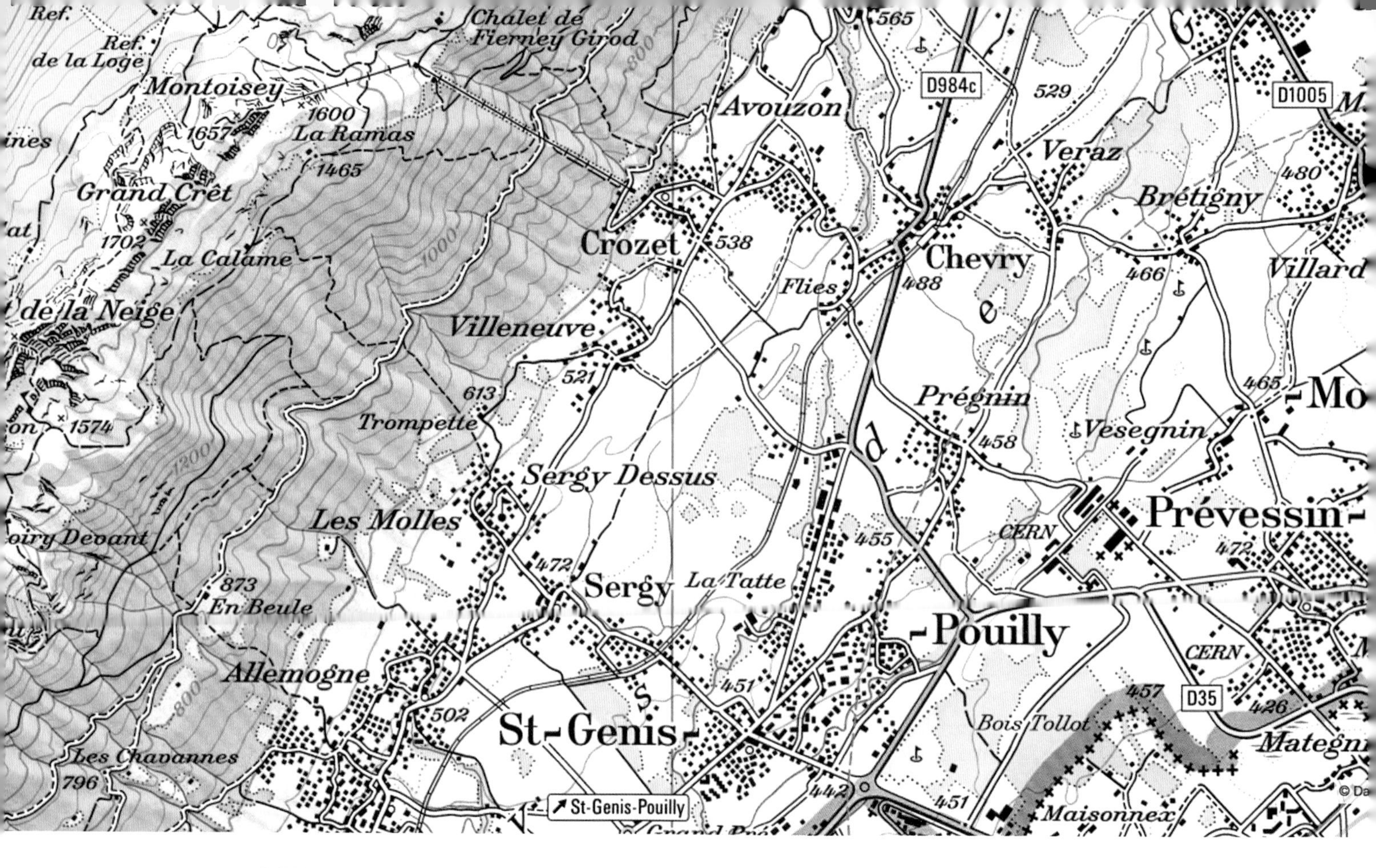

CONTENT-RESPONSIVE LABELS IN HIGH-DIMENSIONAL DATA GRAPHICS

Maps model geography, directly positioning words/numbers/lines/colors/symbols. Topographic maps locate data by latitude/longitude/altitude, 3-space addresses of all geographic elements. Swiss topographic maps show contour data-lines with labels. Direct labels provide exactitude without clutter. Maps also use legends – consistent across all Swiss topographic maps – showing measures of altitude, natural colors for surfaces, and place-names scaled to population size. Good cartography uses multiple layers, accommodating high-dimensional data inputs. Swiss topo maps respect their diverse audience:

trigonometric points	2127.6
spot height	1587 713
index contour	800
lake level	419
spot height lake bottom	387

Town	over 50 000	BERN
Town	10 000 – 50 000	LUGANO
Municipality	2000 – 10 000	Sumvitg
Municipality	less than 2000	Cressier
Suburb	over 2000	*Cassarate*
Suburb	100 – 2000	*Champfèr*
Hamlet, group of houses	50 – 100	*Le Plan*
Single house, hut		*Trifthütte SAC*

1:25 000 1:50 000

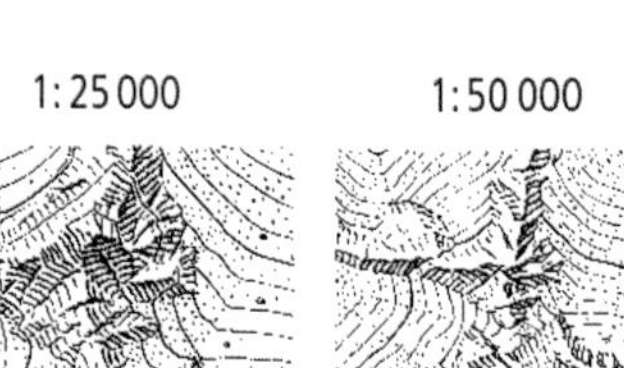

rock with 100 m contour lines

glacier moraine

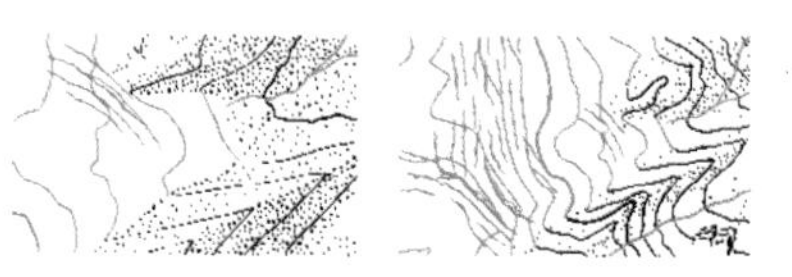

Easily legible map content

Thanks to meticulous reworking, the maps are easy to read. The key changes are:

- easily legible text
- gradation of the road network according to width and colour coding to traffic importance
- red: rail network, railway stations
- larger symbols for legibility
- abandonment of double lines, shaded on one side or dotted, for roads
- closured boundary lines (local cantonal, national borders)

CONTENT-PLACED LABELS/ANNOTATIONS IN A SUPERB, WIDELY-VIEWED DATA GRAPHIC

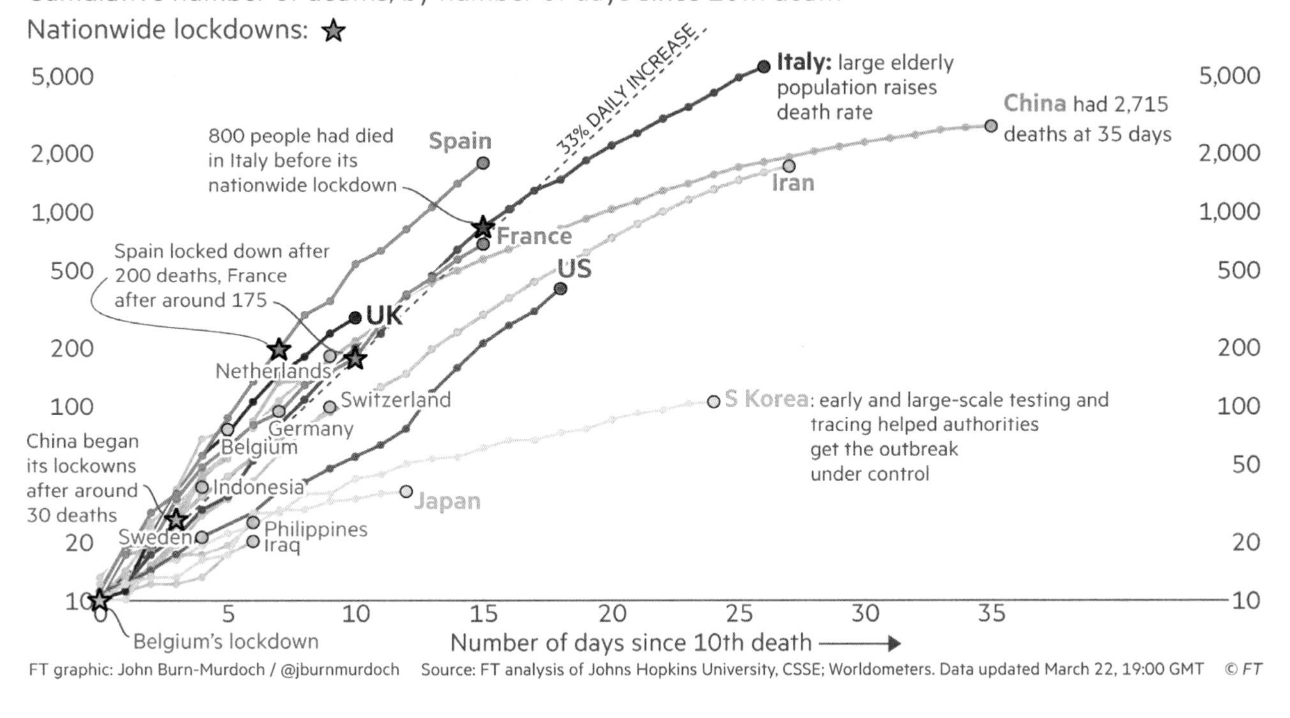

CONVENTIONS ("WE'VE ALWAYS DONE IT THIS WAY") ENSHRINED IN LEGACY CODE CAUSED 50 YEARS OF CONTENT-HOSTILE AND READER-INCONVENIENT DATA GRAPHICS IN POWERPOINT, EXCEL, AND SOPHISTICATED DATA-ANALYSIS COMPUTER PACKAGES

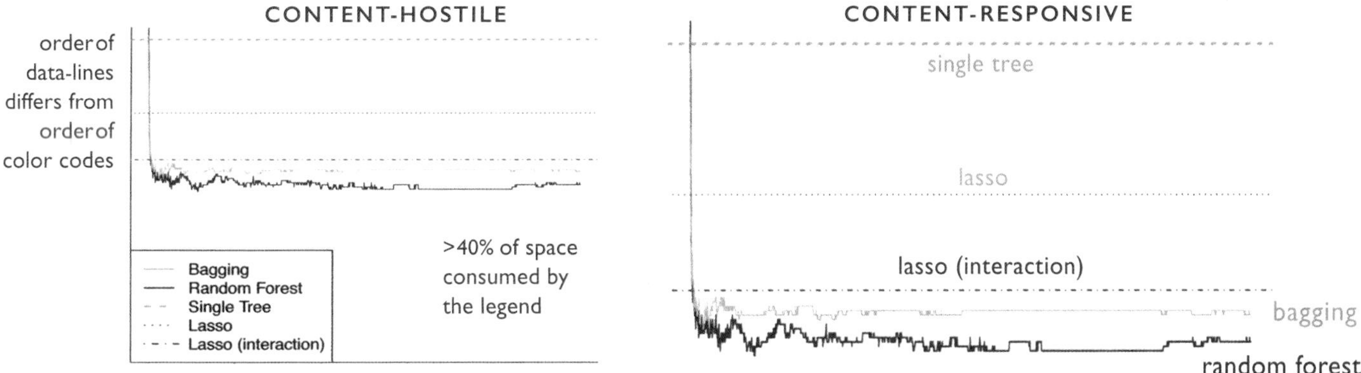

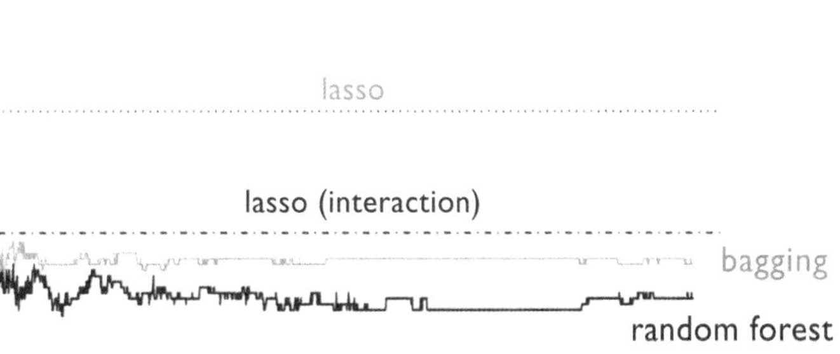

labels should be data-line color, not black

To make sense of this display, readers must briefly memorize a one-time color code stashed in a disordered legend. For 50 years, office and data-analysis software has published trillions of legends – coffins of dead conventions – and trillions of impediments to seeing and learning. *Data graphics should have the same intense commitment to content, clarity, exactitude, integrity as mathematics, maps, computer code, science.* Many more readers look at graphics than at math or code. Statistical inference textbooks use generic data graphics from software packages – and legacy code snippets create endless epidemics of noisome data graphics.

Viewers can read these content-responsive labels *directly,* no encodings. For decades, electronic maps have automated direct labels. Cutting and pasting places content-responsive labels in position. Above, as on maps, the more significant labels are enlarged bagging random forest. All available space shows data. This display is adjusted to show local slopes averaging ~45°. Why should publisher style-sheets or coders decide aspect ratios, placing all data graphics in the same pre-specified box? (See Jeffrey Heer & Maneesh Agrawala, 'Multi-Scale Banking to 45°' *IEEE Transactions on Visualization and Computer Graphics,* 2006, 701-708)

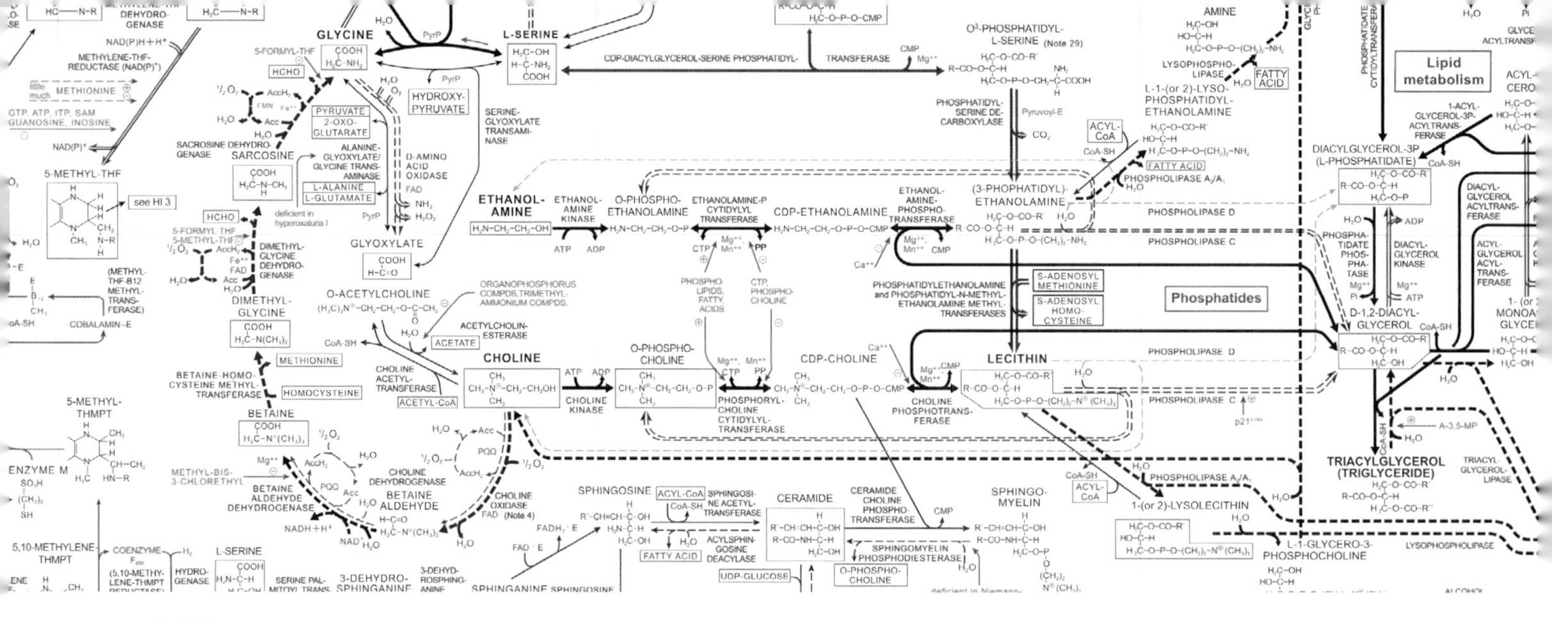

D·6·P·
GLUCONO·1,5-
LACTONE
CH,-O-P
H
H
OH H
O
O
HO
H OH

Low in some
drug hemolytic
anomia types

GLUCOSE 6·P
DEHYDROGENASE

D·GLUCOSAMINE
6·P, higher ACYLCoA
NADPH

NADP·

CONTENT-RESPONSIVE GRAPHICAL SENTENCES DESCRIBE METABOLIC PATHWAYS

Metabolic pathways trace chemical reactions within cells. This display shows 5% of the pathways. Like a map, content is format, a grid of high-resolution meaning. Note use of local annotations. These fundamental structures and processes for many organisms survived evolutionary stress:

> "A striking feature of metabolism is the similarity of the basic metabolic pathways and components between even vastly different species. For example, the set of carboxylic acids that are best known as intermediates in the citric acid cycle are present in all known organisms, being found in species as diverse as the unicellular bacterium *Escherichia coli* and huge multicellular organisms like elephants. These striking similarities in metabolic pathways are likely due to their early appearance in evolutionary history, and their retention because of their efficacy." WIKIPEDIA

CONTENT-RESPONSIVE GRAPHICAL SENTENCE NARRATES SEQUENCE OF DATA COLLECTION

With several stacklists, this sentence reports 22 numbers describing participants in a randomized controlled trial comparing blood pressure reduction by intensive treatment goal (120 systolic) compared to standard treatment goal (140 systolic):

14,692 patients were assessed for eligibility → 5,331 were ineligible or declined to participate:

- 34 were less than 50 years of age
- 352 had low systolic blood pressure
- 2,284 were taking too many medications
- 718 were not at increased cardiovascular risk
- 703 had miscellaneous reasons
- 587 did not give consent
- 653 did not complete screening

9,361 were eligible, underwent random assignment

4,678 randomly assigned to intensive treatment
- 224 discontinued intervention
- 111 were lost to follow-up
- 154 withdrew consent

4,678 analyzed (includes all intent to treat) intensive treatment of blood pressure with goal of 120 systolic

4,683 randomly assigned to standard treatment
- 242 discontinued intervention
- 134 were lost to follow-up
- 121 withdrew consent

4,683 analyzed (includes all intent to treat) standard treatment of high blood pressure with goal of 140 systolic

SMALL MOVES, FINE DISTINCTIONS, SOME CONSEQUENCES

CONTENT-INDIFFERENT LABELS IN DATA GRAPHICS CREATE NOISE IN BILLIONS OF PAGE VIEWS

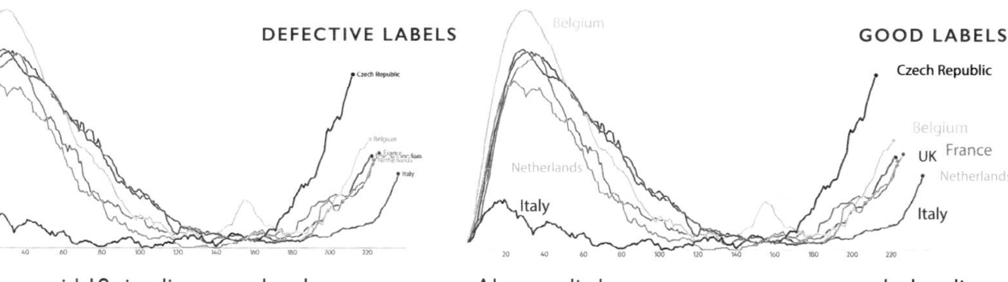

In many covid-19 timelines, graphs place country names in tiny type at far right. At first glance, especially on small display screens, lines appear to suddenly turn flat for the most recent data. Data-lines and country-names are too close together, optically merging to create false news.

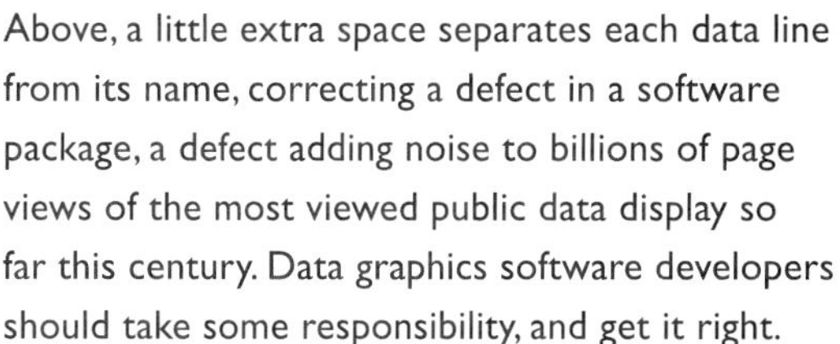

Above, a little extra space separates each data line from its name, correcting a defect in a software package, a defect adding noise to billions of page views of the most viewed public data display so far this century. Data graphics software developers should take some responsibility, and get it right.

TO ORGANIZE DIALOGUE, NOVELS USE CONTENT-RESPONSIVE PARAGRAPH INDENTS

"When *I* use a word," Humpty Dumpty said in rather a scornful tone, "it means just what I choose it to mean — neither more nor less."

"The question is," said Alice, "whether you *can* make words mean so many different things."

"The question is," said Humpty Dumpty, "which is to be master — that's all." LEWIS CARROLL THROUGH THE LOOKING GLASS

SLOWING DOWN THE SENTENCE

"Elmore Leonard possesses gifts – of ear and eye, of timing and phrasing – that even the most snobbish masters of the mainstream must vigorously covet. And the question is: how does he allow these gifts to play, in his efficient, unpretentious and (delightfully) similar yarns about semiliterate hustlers, mobsters, go-go dancers, cocktail waitresses, loan sharks, bounty hunters, blackmailers, and syndicate executioners? My answer may sound reductive, but here goes: the essence of Elmore is to be found is his use of the present participle.

What this means, in effect, is that he has discovered a way of slowing down and suspending the English sentence – or let's say the American sentence, because Mr. Leonard is as American as jazz. Instead of writing 'Warren Ganz III lived up in Manalapan, Palm Beach County,' Mr. Leonard writes: 'Warren Ganz III, living up in Manalapan, Palm Beach County.' He writes, 'Bobby saying,' then opens quotes. He writes, 'Dawn saying,' and then opens quotes. We are not in the imperfect tense (Dawn was saying) or the present (Dawn says) or the historic present (Dawn said). We are in a kind of marijuana tense (Dawn saying), creamy, wandering, weak-verbed. Such seem to open up a lag in time, through which Mr. Leonard easily slides, gaining entry to his players' hidden minds. He doesn't just show you what these people say and do. He shows you where they breathe."

MARTIN AMIS 'ON ELMORE LEONARD' THE WAR AGAINST CLICHÉ

MINIFY HYPHENS, WHICH HAVE TINY MEANING

not this: e–mail
use this: e-mail
evolving to: email

MAXIFY DASHES, THEIR MEANING IS OFTEN AS STRONG AS LINE-BREAKS IN POETRY

'Elmore Leonard possesses gifts –
of ear and eye, of timing and phrasing –
that even the most snobbish masters of
the mainstream must vigorously covet.'

TINY PAUSES, SILENCES, AND SPACES REMODEL MUSIC

Silence is the basis of music.
We find it before, after, in, underneath, behind the sound.
Some pieces emerge out of silence or lead back into it.

ALFRED BRENDEL

Listening to a recording of Beethoven's Third Piano Concerto, Haruki Murakami and Seiji Ozawa celebrate Glenn Gould's control of pauses, silences, empty spaces:

"In the recording, the woodwinds play, and Gould adds his arpeggios:

OZAWA 'Here it is – the part you were talking about.'

MURAKAMI 'Yes, this is it. The piano is supposed to be accompanying the orchestra, but Gould's touch is so clear and deliberate.'

Gould ends a phrase, takes a brief pause, moves on to the next phrase.

OZAWA 'Now that – where he took that pause – that's absolutely Glenn at his freest. It's the hallmark of his style, those perfectly timed empty spaces.'

The piano and orchestra intertwine beautifully for a while.

OZAWA 'Now, we're completely in Glenn Gould's world. In Japan we talk about *ma* in Asian music – the importance of pauses or empty spaces – but it's there in Western music, too. You get a musician like Glenn Gould, and he's doing exactly the same thing. Not everybody can do it – certainly no ordinary musician. But somebody like him does it all the time.'

MURAKAMI 'No, never. Or if they do, the spaces don't fit in as naturally as this. It doesn't grab you – you don't get drawn in as you do here. That's what putting in these empty spaces, or *ma*, is all about. You grab your audience and pull them in. East or West, it's all the same when a virtuoso does it.'"

HARUKI MURAKAMI AND SEIJI OZAWA, ABSOLUTELY ON MUSIC: CONVERSATIONS, 2016

FINE DISTINCTIONS ACCUMULATE INTO MEANING

The *Große Fuge* is complex, and Beethoven instructs the quartet to play *sempre piano* – always softly gently. Leonard Ratner on this section: 'a wonderful change of color, offered with the silkiest of textures, and with exquisite moments of glowing diatonism.'

Wikipedia, *Große Fuge*, edited.

REMODELING TEXT-ONLY PARAGRAPHS: MULTIMODAL PARAGRAPHS ASSIST REMEMBERING AND LEARNING

MULTIPLE MODES OF INFORMATION Explaining, thinking, learning, remembering are inherently multimodal – words, numbers, graphics, images, symbols, models, whatever it takes. Avoid segregating information by mode of production. Meaning and truth are resolved by all relevant evidence, regardless of form, format, mode.

MULTIMODAL PARAGRAPHS: UNIQUE, MEMORABLE, CLOSER TO TRUTH Text-only paragraphs are typographically identical – typeface, spacings, line-lengths all aligned down deep columns. Authors and content have no control over space. This locked-down, boring regularity of text paragraphs makes design and production convenient. But inconvenient for readers, who are unable to find and read again a specific string of words in previously-read paragraphs. Most readers encounter this dilemma in essays, news reports, research articles, novels. By uniquely activating the relevant neural substrates for retaining visual memories, multimodal paragraphs assist memory, retrieval, and understanding by readers – as in dictionaries with illustrations. Nearly every paragraph in this book is content-driven and deliberately visually unique.

BILLIONS OF EXAMPLES OF CONTENT-RESPONSIVE MULTIMODAL PARAGRAPHS

THE WHOLE EARTH CATALOG (1970) = 2000 CONTENT-DRIVEN MULTIMODAL PARAGRAPHS

A VISUAL INDEX OF THIS BOOK (CHAPTER 9): 200 MULTIMODAL PARAGRAPHS

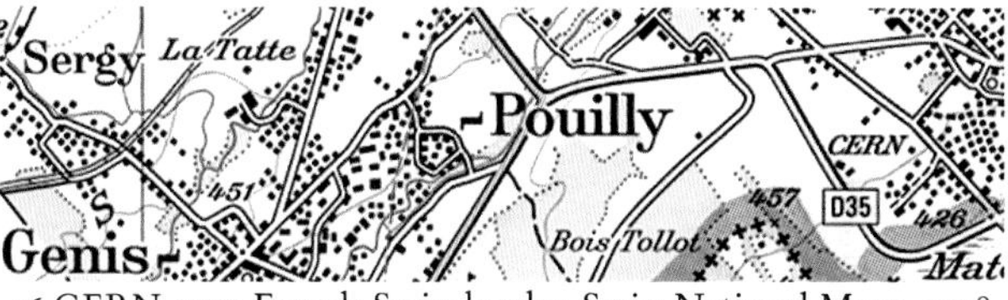

56 CERN, near French-Swiss border. Swiss National Maps, 2018,

27 Yehudi Menuhin's marked-up copy of J. S. Bach's *Solo Violin Sonata No. 2 in A minor*, 1720, Foyle Menuhin Archive

ALDUS MANUTIUS, HYPNEROTOMACHIA (1499)

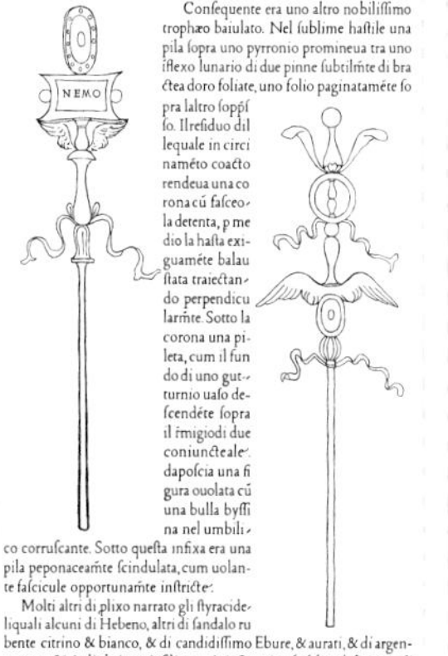

Conſequente era uno altro nobiliſſimo trophæo baiulato. Nel ſublime haſtile una pila ſopra uno pyrronio promineua tra uno iflexo lunario di due pinne ſubtilmte di bractea doro foliate, uno folio paginataméte ſopra laltro ſoppſſo. Il reſiduo dil lequale in circinaméto coacto rendeua una corona cú faſceola detenta, p medio la haſta exiguaméte balauſtata traiectando perpendicularmte. Sotto la corona una pileta, cum il fundo di uno gutturnio uaſo deſcendéte ſopra il ſmigiodi due coniuncteale. dapoſcia una figura ouolata cú una bulla byſſina nel umbilico corruſcante. Sotto queſta infixa era una pila peponaceamte ſcindulata, cum uolante faſcicule opportunamte inſtricte.

Molti altri di plixo narrato gli ſtyracideliquali alcuni di Hebeno, altri di ſandalo rubente citrino & bianco, & di candidiſſimo Ebure, & aurati, & di argento contecti, & di altri pretioſilignamini. Omni coſa fabre deformata di tenuiſſimo oro, argento, & di leuigata materia, & di ſeta uirente formati & di omni altra gratioſa coloratione, cum iucunda floratura. Cum gemme multiplicemente ornati agli congruenti lochi omni coſa harmonicamé te deſtinata & conſpicuamente applicata, cum præpendenti ſpondili, o-

x

palmiqtro. Nelqle in una facie, dal frõte dila fractura era iſcripto, & ſimilméte oue era rupto pidicio di alcune litere pte fragmétate, & itegre, parte rimaſte. Poſcia nella ſubiecta corpulétia dalla circinante cinctura uerſo el fondo, nellaquale erano appacte le anſe, nel fronte dilla fractura era queſta præſtante ſcriptura.

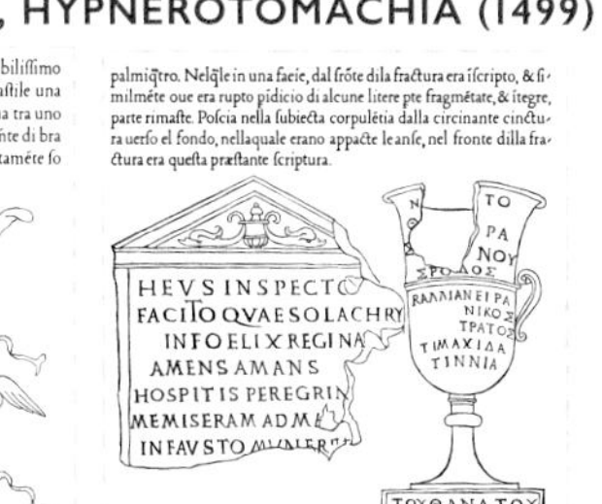

Relicti queſti rupti monumenti, ad una deſtructa tribuna deueni nellaquale alquanto fragmento di muſeaco ſi comprendeua. Oue picto mirai uno homo affligente una damicella. Et uno naufragio. Et uno adoleſcétulo ſopra il ſuo dorſo equitante una fanciulla, nataua ad uno littore deſerto. Et parte uedeuaſi di uno leone. Et quegli dui in una nauicula remiganti. Il ſequente diſtructo. Et ancora queſta parte era in molti lochi lacerata, Non ualeua intendere totalmente la hiſtoria. Ma nel pariete cruſtato marmoreo, era interſepta una tabula ænea, cum maiuſcule græcæ. Tale epigramma inſcripto hauea. Ilquale nel proprio Idiomate in tãta pietate me prouocaua legendo ſi miſerando caſo, che di lachryme contenirme non potui, dánando la rea fortuna. Ilquale ſæpicule perlegendo, quanto io ho potuto cuſi il fece latino.

SOME GOOD TEXT-IMAGE INTEGRATION IN TWEETS, BUT HUNDREDS OF MILLIONS OF USERS/VIEWERS ENCOUNTER IMPOVERISHED TYPOGRAPHY

News & Views: A Nature paper reports an accessible machine-learning tool that can accelerate the optimization of a wide range of synthetic reactions — and reveals how cognitive bias might have undermined optimization by humans.

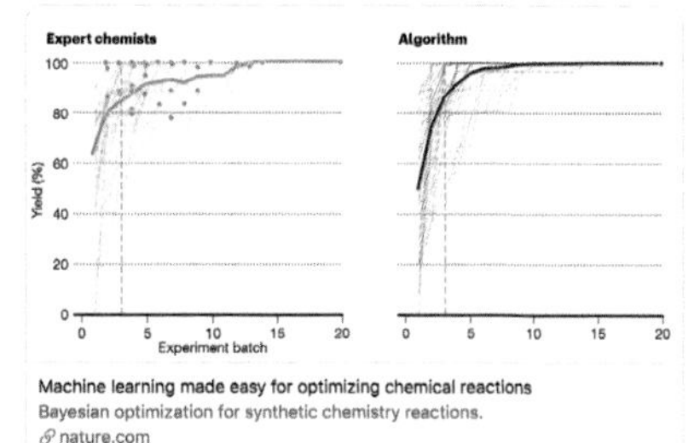

Expert chemists
Algorithm
Yield (%)
Experiment batch
Machine learning made easy for optimizing chemical reactions
Bayesian optimization for synthetic chemistry reactions.
nature.com

Megan Jaegerman's brilliant news graphics, a portfolio of about 40 different graphics. edwardtufte.com/bboard/q-and-a... #GraphicDesign #ddj #journalism

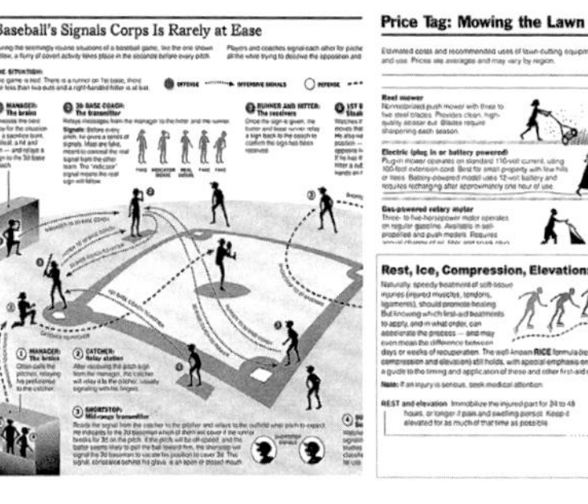

San Francisco International Airport (KSFO)

lat 37.62°N long -122.37°W elevation 10 feet (3 m)

A few clouds
78°F
26°C
Humidity 39%
Wind speed N 5 mph (N 8 kmph)
Barometer 30.08 in (1018.6 mb)
Dewpoint 51°F (11°C)
Visibility 10 miles (16 km)
Heat index 79°F (26°C)
Last update 16 April 1:56 pm PDT

Forecast office More local weather 3 day history Hourly weather forecast

MULTIMODAL DATA PARAGRAPHS = WORDS, NUMBERS, TEXT-TABLES, IMAGES

Viewed by millions everyday, National Weather Service data paragraphs describe current weather: 18 measurements (U.S./metric), 28 words, 4 links to weather data, and a sunny illustration for users who can't look out the window. This table also serves weather experts interested in the 5th significant digit of barometric pressure. Good displays can *simultaneously serve multiple interests and constituencies.* Presentations should make everyone smarter, not merely dumb things down.

Think of the data paragraph as the fundamental unit of presentation. Show small data sets directly, as in this data paragraph. Avoid Little Data Graphics. Pie charts, bar charts encode numbers into areas/colors. Viewers then retranslate areas/colors back into numbers. Many encodings are unique, used only once, and thus do not repay learning. Little Data Graphics are for pitches, not truth. Data visualizations are at their best when there is so much data that the only way to see it . . . is to see it.

Donald Knuth, using his layout and typographic tools TEX and METAFONT, published ~10^4 multimodal paragraphs in 30 books integrating words, code, illustrations, images, data graphics.

A high-resolution inline data graphic, a skyline sparkline: a letter-form box, 100 pt wide x 10 pt tall = 1000 locations with addresses = a data-graphic.

A random "skyline" texture, 100 pt wide × 10 pt tall: The density decreases uniformly as you go up in altitude.

Here, 20 scatterplots work at typographic resolution (like sparklines), showing 4,000 numbers. Scatterplots are constructed in a letter-form grid. Code, like the hand, can be indifferent to modes of information.

The resulting "characters," if we repeat the experiment ten times, look like this:

And if we replace 'uniformdeviate w' by '$.5w + w/6 *$ normaldeviate', we get

THE WORLD WE SEEK TO UNDERSTAND IS MULTIMODAL AND MULTIVARIATE

In the beginning was not the word.
The book of Nature is written in mathematics, not words.
The book of life is written in DNA, not words.
Scientific certainty is visible empirical evidence, not wordy authority.

In the beginning of the computer age was the proprietary format, which messed up content/meaning/space, stole ownership of content, segregated modes of information. To understand Nature's laws, biological systems, human behavior requires all modes of information gathered together, just as our perception of the world simultaneously detects and reasons about many different modes of information.

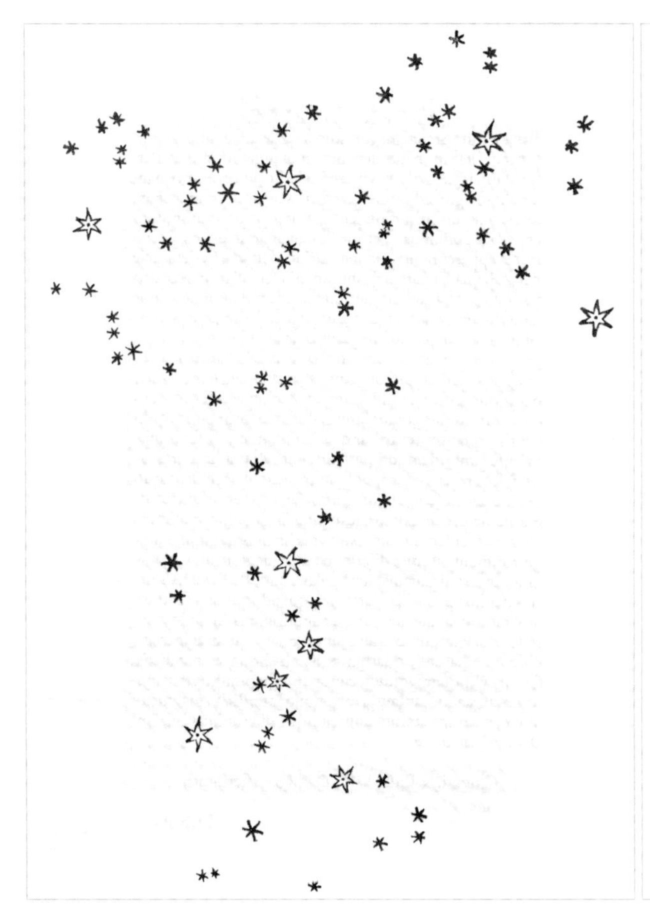

Quòd tertio loco à nobis fuit obſeruatum, eſt ipſiuſmet LACTEI Circuli eſſentia, ſeu materies, quam Perſpicilli beneficio adeò ad ſenſum licet intueri, vt & altercationes omnes, quæ per tot ſæcula Philoſophos excruciarunt ab oculata certitudine dirimantur, noſque à verboſis diſputationibus liberemur. Eſt enim GALAXYA nihil aliud, quam innumerarum Stellarum coaceruatim conſitarum congeries; in quamcunq; enim regionem illius Perſpicillum dirigas, ſtatim Stellarum ingens frequentia ſe ſe in conſpectum profert, quarum complures ſatis magnæ, ac valde conſpicuæ videntur; ſed exiguarum multitudo prorſus inexplorabilis eſt.

At cum non tantum in GALAXYA lacteus ille candor, veluti albicantis nubis ſpectetur, ſed complures conſimilis coloris areolæ ſparſim per æthera ſubfulgeant, ſi in illarum quamlibet Specillum conuertas Stellarum conſtipatarum

cętum

VISIBLE CERTAINTY, THE EVIDENCE OF THE EYE, LIBERATES US FROM WORDY ARGUMENTS

Galileo's *Sidereus Nuncius* (1610) shows the visible stars ✩ seen by unaided eyes, as well as innumerable 'invisible stars' now seen by his telescope * ✱ ✱ Stars flow into the margins and off the pages into an unbounded cosmic vastness, *breaking out of the typographic grid, for Nature is nowhere rectangular. And nowhere flat, except for a perfectly still pond.*

Galileo Galilei, *Sidereus Nuncius* (Venice, 1610), middle 2 pages of the unpaginated 4-page insert between pages 16V and 17R.

Galileo sharply distinguishes between word and image in reasoning about celestial objects. Before 1610, astronomy was verbal gymnastics, speculation, philosophizing, parsing Aristotle and sacred texts. In contrast, the new telescopic images are the direct, visible, decisive testimony of Nature herself. Galileo tied empirical observation to truth with the words oculata certitudine, visible certainty. From then on, all science, to be credible, depended on publicly displayed evidence of seeing/reasoning, not words. Galileo wrote:

Revised from Edward Tufte, *Beautiful Evidence* (2006), 96-109.

> 'What was observed is the nature or matter of the Milky Way itself, which, with the aid of the spyglass, may be observed so well that all the disputes that for so many generations have vexed philosophers are destroyed by visible certainty, and we are liberated from wordy arguments. For the Galaxy is nothing else than a congeries of innumerable stars distributed in clusters.
> To whatever region you direct your spyglass, an immense number of stars immediately offer themselves to view, of which very many appear rather large and very conspicuous but the multitude of small ones is truly unfathomable.'

Sidereus Nuncius, translated by Albert Van Helden (1989), 62.

3 GRAPHICAL SENTENCES: NOUNS AND VERBS, STRUCTURE AND FUNCTION

In ~1560 Ettore Ausonio, a polymath with interests from mathematics to mirror-making, constructed this immense (44 x 74 cm) diagram depicting reflections from concave spherical mirrors. Then, between 1592 and 1601, while teaching at the University of Padua, Galileo made this handwritten copy of the diagram, which was fortunate since Ausonio's original has since gone missing. Three helpful architectures for off-the-grid sentences are deployed – *word trees, stacklists, annotated linking lines:*

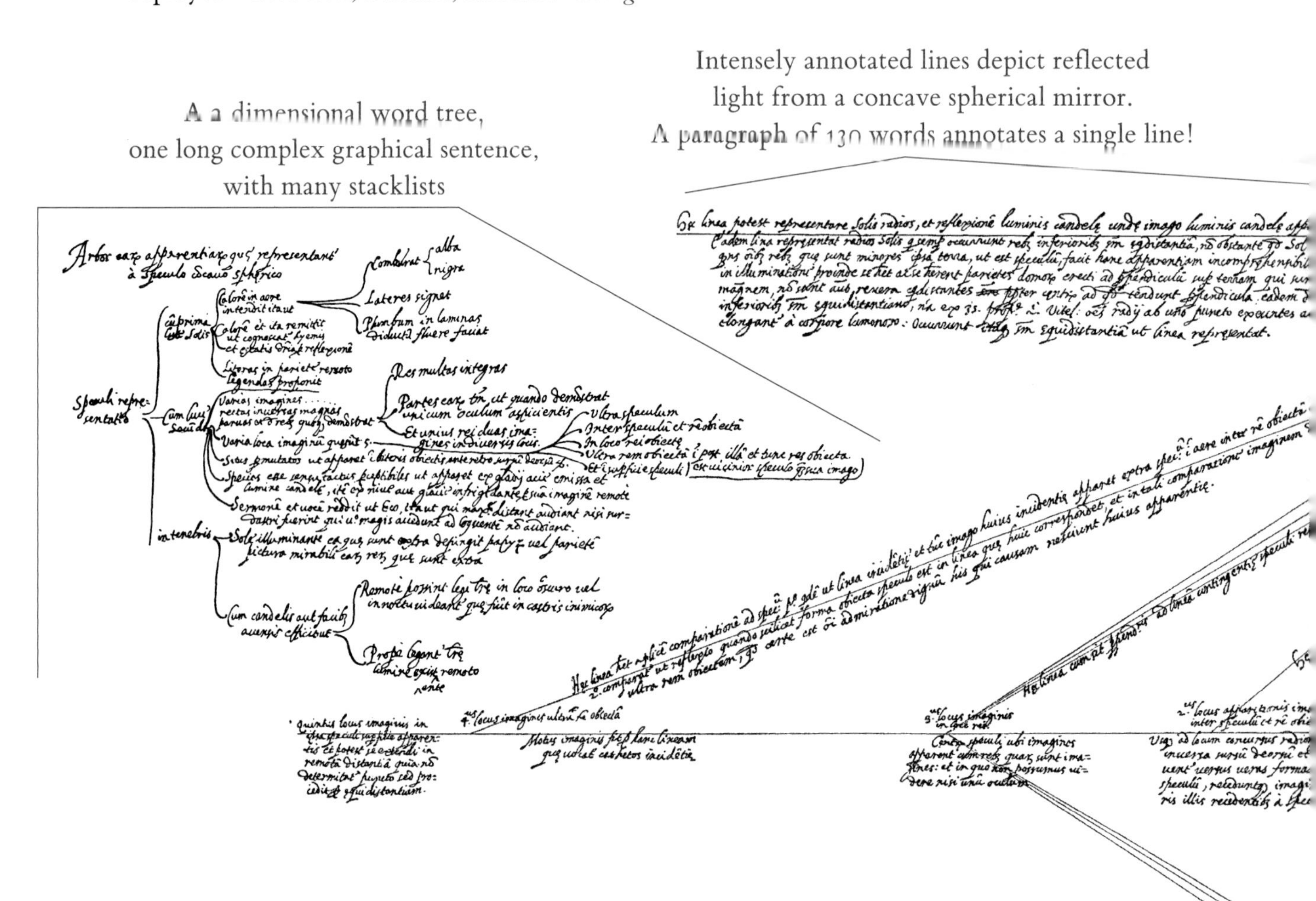

Intensely annotated lines depict reflected light from a concave spherical mirror. A paragraph of 130 words annotates a single line!
A dimensional word tree, one long complex graphical sentence, with many stacklists

GRAPHICAL SENTENCES LIVE IN THE PLANE, NOT IN THE LINE

FIERCELY CONTENT-RESPONSIVE HANDWRITING
BREAKS CONVENTIONAL TYPOGRAPHIC GRIDS.
HERE, THE IMAGE IS THE GRID.

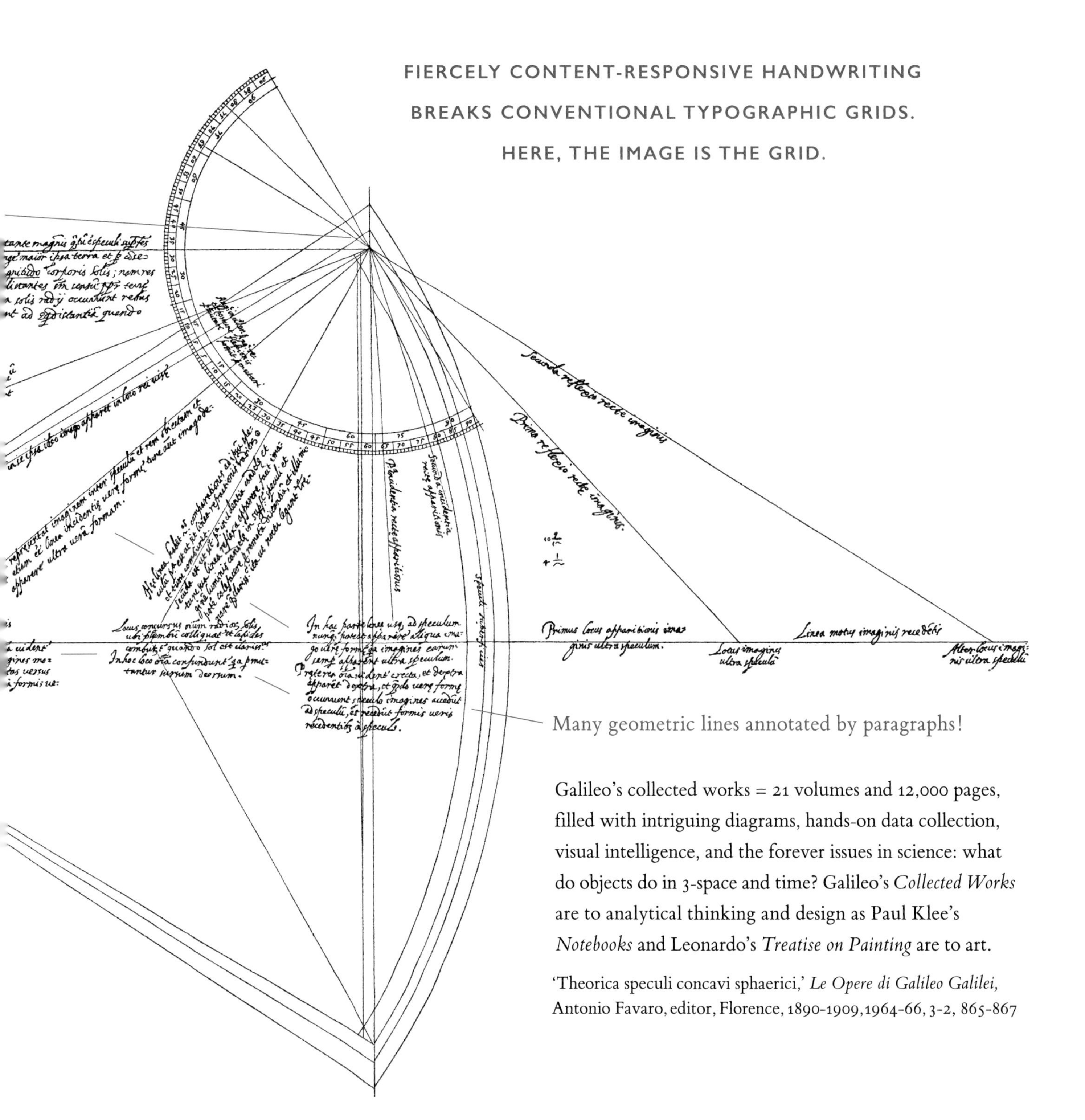

Many geometric lines annotated by paragraphs!

Galileo's collected works = 21 volumes and 12,000 pages, filled with intriguing diagrams, hands-on data collection, visual intelligence, and the forever issues in science: what do objects do in 3-space and time? Galileo's *Collected Works* are to analytical thinking and design as Paul Klee's *Notebooks* and Leonardo's *Treatise on Painting* are to art.

'Theorica speculi concavi sphaerici,' *Le Opere di Galileo Galilei,* Antonio Favaro, editor, Florence, 1890-1909, 1964-66, 3-2, 865-867

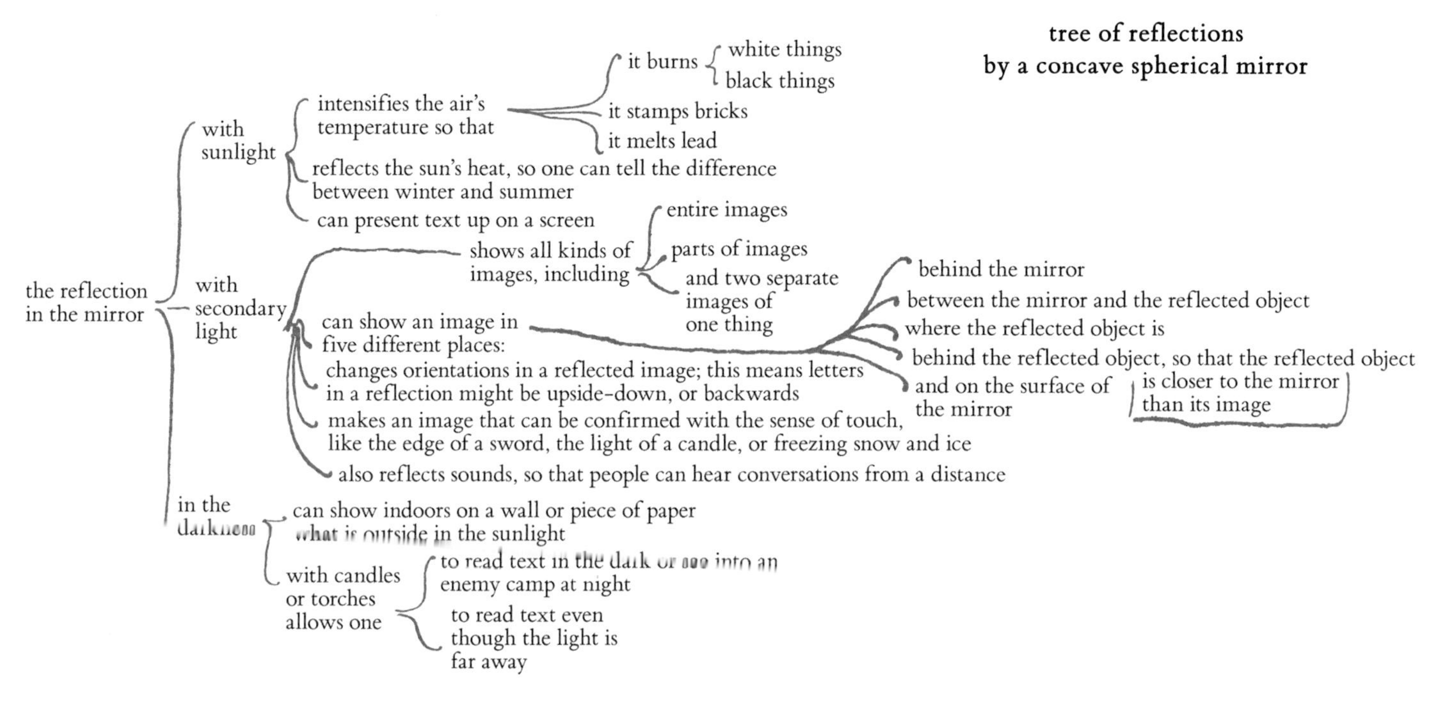

Tree of Reflections by a Concave Spherical Mirror is a circus, a 2-dimensional graphical sentence with 218 words, 29 branches, 13 stacklists. Unlike the more rigorous logic of Markov chains and decision trees, its branches meander. This word architecture is intriguing, different – intensely annotated lines and *stacklists*, their branches ending in full leaf, quasi-paragraphs of words.

Here is my graphical sentence with 2 stacklists:

<table>
<tr><td>Ausonio's Tree of Reflections suggests a useful</td><td>architecture
stack
word tree
design
typography
haystack</td><td>for displaying</td><td>lists
taxonomies
cladograms
flows
synonyms
thesauruses
branching data
whatever</td></tr>
</table>

Stacklists organize and clarify complex material in 2-space. Readers read more slowly, and that's good: to think, look again, and connect words *vertically within each stack and horizontally between stacks*. Instead of polyphony, conventional inline lists are a freight train of words along a one-way narrow track, making it difficult to identify which words belong to which list and to link/compare elements within/between lists:

> Ausonio's *Tree of Reflections* suggests a useful architecture, stack, word tree, design, typography, haystack for displaying lists, taxonomies, cladograms, flows, synonyms, thesauruses, branching data, whatever.

A GRAPHICAL SENTENCE: FEYNMAN DIAGRAMS DEPICT SUBATOMIC PARTICLES IN ACTION.

THIS FEYNMAN DIAGRAM IS ALSO A TYPOGRAPHIC GRID, TELLING THE WORDS WHERE THEY BELONG.

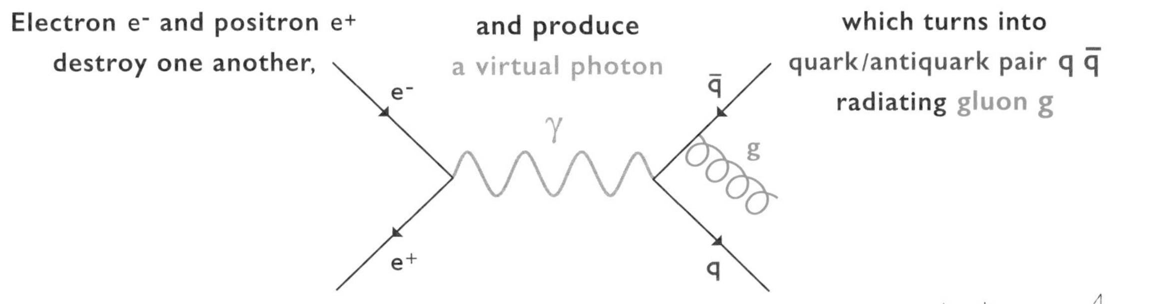
Electron e⁻ and positron e+
destroy one another,
and produce
a virtual photon
which turns into
quark/antiquark pair q q̄
radiating gluon g
e⁻
e+
γ
q̄
g
q

Perhaps diagrams of universals are universal, since some diagrams portray Nature's universal processes – such as metabolic pathways, Feynman diagrams, and hyperfine transitions of neutral hydrogen (as shown in a diagram mounted on the Pioneer Space Plaque for potential extra-terrestrial readers).

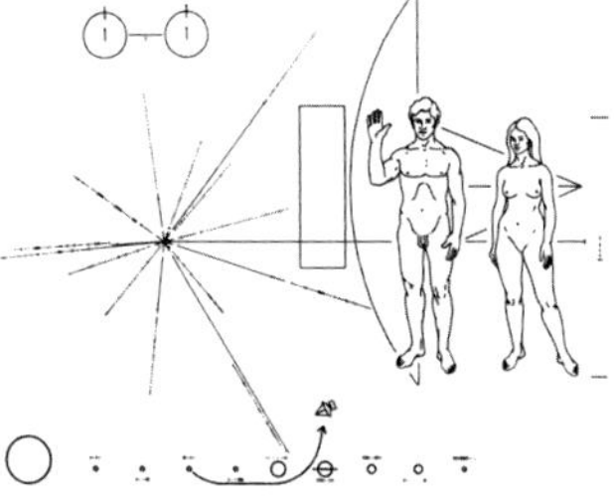

STACKLISTS CLARIFY COMPLEX MEANING, ENCOURAGE SLOW, THOUGHTFUL READING

All day long	mapmakers illustrators painters typographers architects sculptors tectonic plates	construct and contrive interactions at borderlines and borderspaces	that shape space and enhance meaning create real/apparent surfaces and volumes strengthen internal unity of fields separate inside from outside.

This graphical sentence deploys 11 line-breaks, 3 stacklists, 40 words.

Reading a graphical sentence requires no instruction manual, no previous experience.

This sentence makes broad substantive claims, and the spacing encourages slow, thoughtful reading. The idea is to maximize explanatory depth, not speed-reading.

How meaning and spacing work in this graphical sentence:

All day long	mapmakers illustrators painters typographers architects sculptors tectonic plates	construct and contrive interactions at borderlines and borderspaces	that shape space and enhance meaning create real/apparent surfaces and volumes strengthen internal unity of fields separate inside from outside.

Meaning Fullness of time announced at beginning of sentence	Meaning Subjects of this sentence: a broad, complex set Spacing 7 list-elements, with content-responsive spaces.	Meaning All the different subjects, from mapmakers to tectonic plates, construct and contrive interactions Spacing Stacklist separating verbs, verb objects, location of actions	Meaning Five verb clauses describe what is done all day long by the sentence subjects Spacing Accented in a stack list, verb clauses reside in a plane, not in a line. Central-axis typography isolates each verb clause

37 GRAPHICAL SENTENCES: NOUNS AND VERBS, STRUCTURE AND FUNCTION

Linking lines aspire to be sentences – with subjects, verbs, objects. Annotated links show *what things do:* causes and effects, mechanisms, trade-offs, flows, feedback, interactions. Unannotated links lack character. No more $x - y$ or $x \rightarrow y$ saying that x and y have something, who-knows-what, to do with one another. Showing both structure and function, 37 graphical sentences in the Walt Disney Productions 1957 profit formula reveal a lively complex network of interactions and job instructions. This amazing diagram demonstrates *real worlds do not consist of anonymous one-way streets,* as assumed in organization charts with their generic know-nothing links between boxes filled with proper nouns.

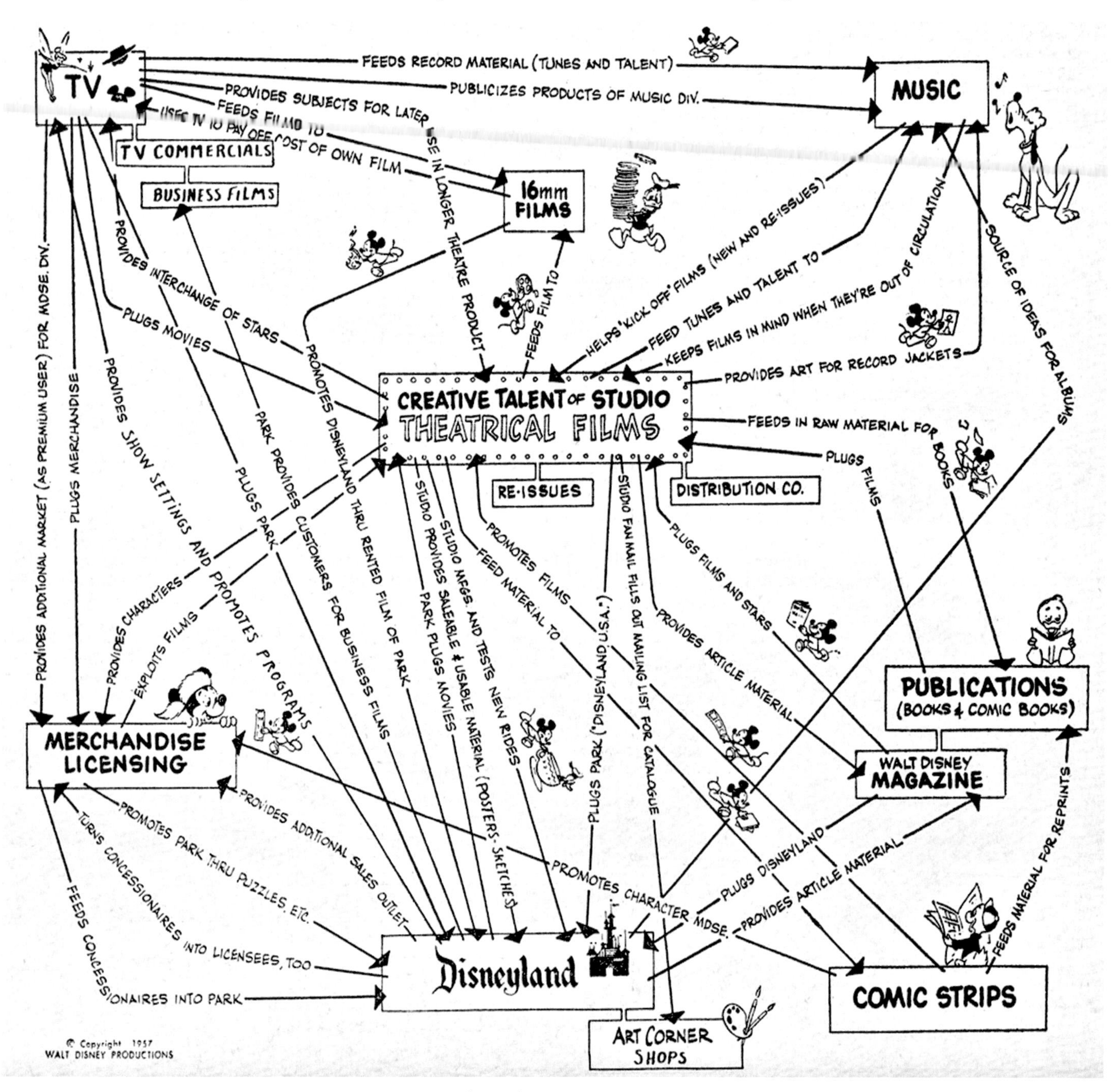

Drawn by Walt Disney, © 1957 Disney

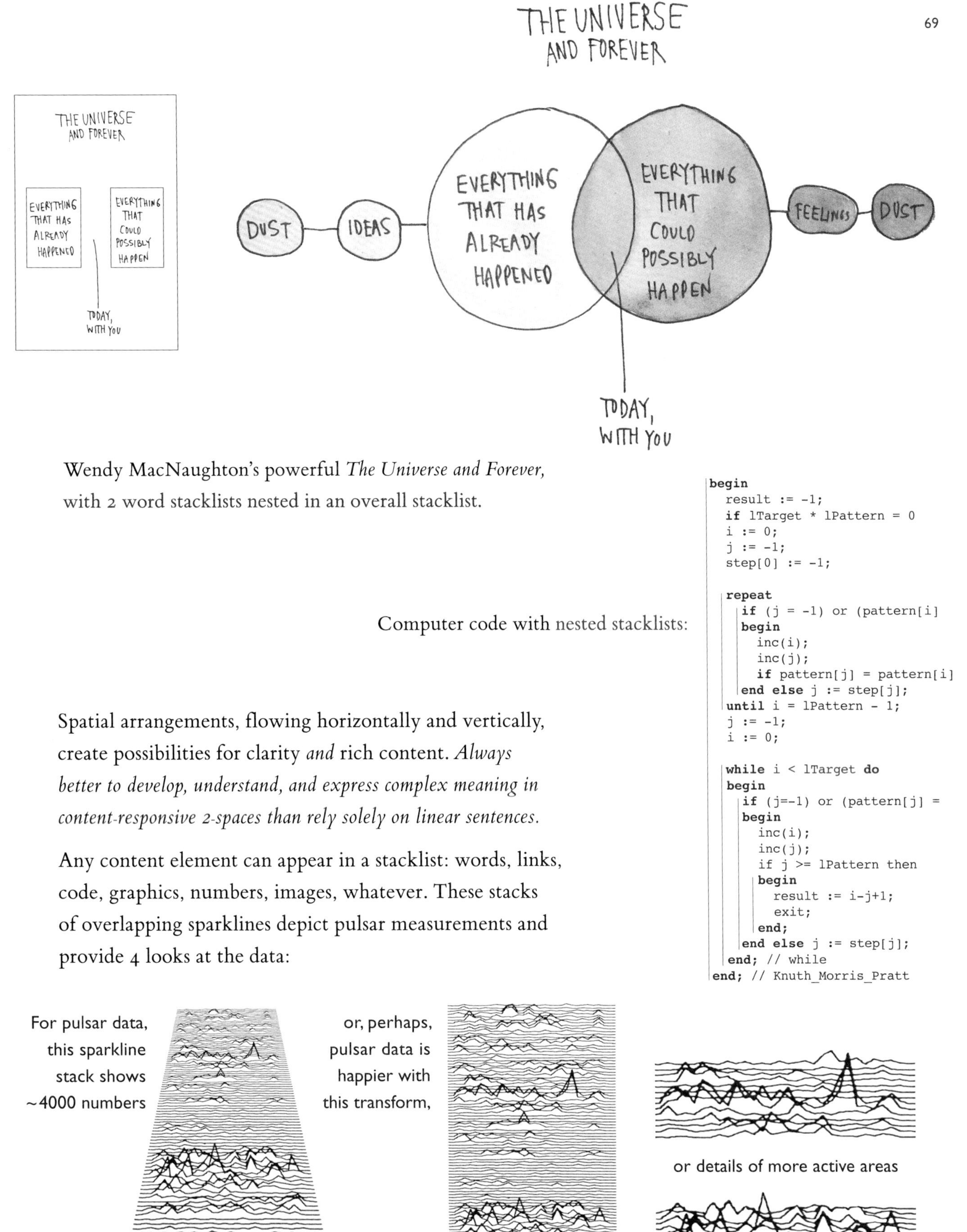

Wendy MacNaughton's powerful *The Universe and Forever,* with 2 word stacklists nested in an overall stacklist.

Computer code with nested stacklists:

```
begin
  result := -1;
  if lTarget * lPattern = 0
  i := 0;
  j := -1;
  step[0] := -1;

  repeat
    if (j = -1) or (pattern[i]
    begin
      inc(i);
      inc(j);
      if pattern[j] = pattern[i]
    end else j := step[j];
  until i = lPattern - 1;
  j := -1;
  i := 0;

  while i < lTarget do
  begin
    if (j=-1) or (pattern[j] =
    begin
      inc(i);
      inc(j);
      if j >= lPattern then
      begin
        result := i-j+1;
        exit;
      end;
    end else j := step[j];
  end; // while
end; // Knuth_Morris_Pratt
```

Spatial arrangements, flowing horizontally and vertically, create possibilities for clarity *and* rich content. *Always better to develop, understand, and express complex meaning in content-responsive 2-spaces than rely solely on linear sentences.*

Any content element can appear in a stacklist: words, links, code, graphics, numbers, images, whatever. These stacks of overlapping sparklines depict pulsar measurements and provide 4 looks at the data:

For pulsar data, this sparkline stack shows ~4000 numbers

or, perhaps, pulsar data is happier with this transform,

or details of more active areas

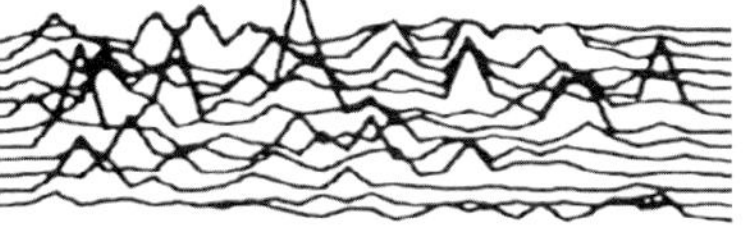

Linked stacklists of 6 black words and 10 red words = 120 word pairs. Brute force heuristics for creative thinking juxtapose familiar words in unfamiliar ways. Escaping the glut of all possible n! combinatorial excesses requires a lively sense of the relevant.

Linking lines between stacklists depict readings that are multiple, alternative, walk-through, improvisational.

Linked stacks show the logic of code sequence and operations, by the great Donald Knuth:

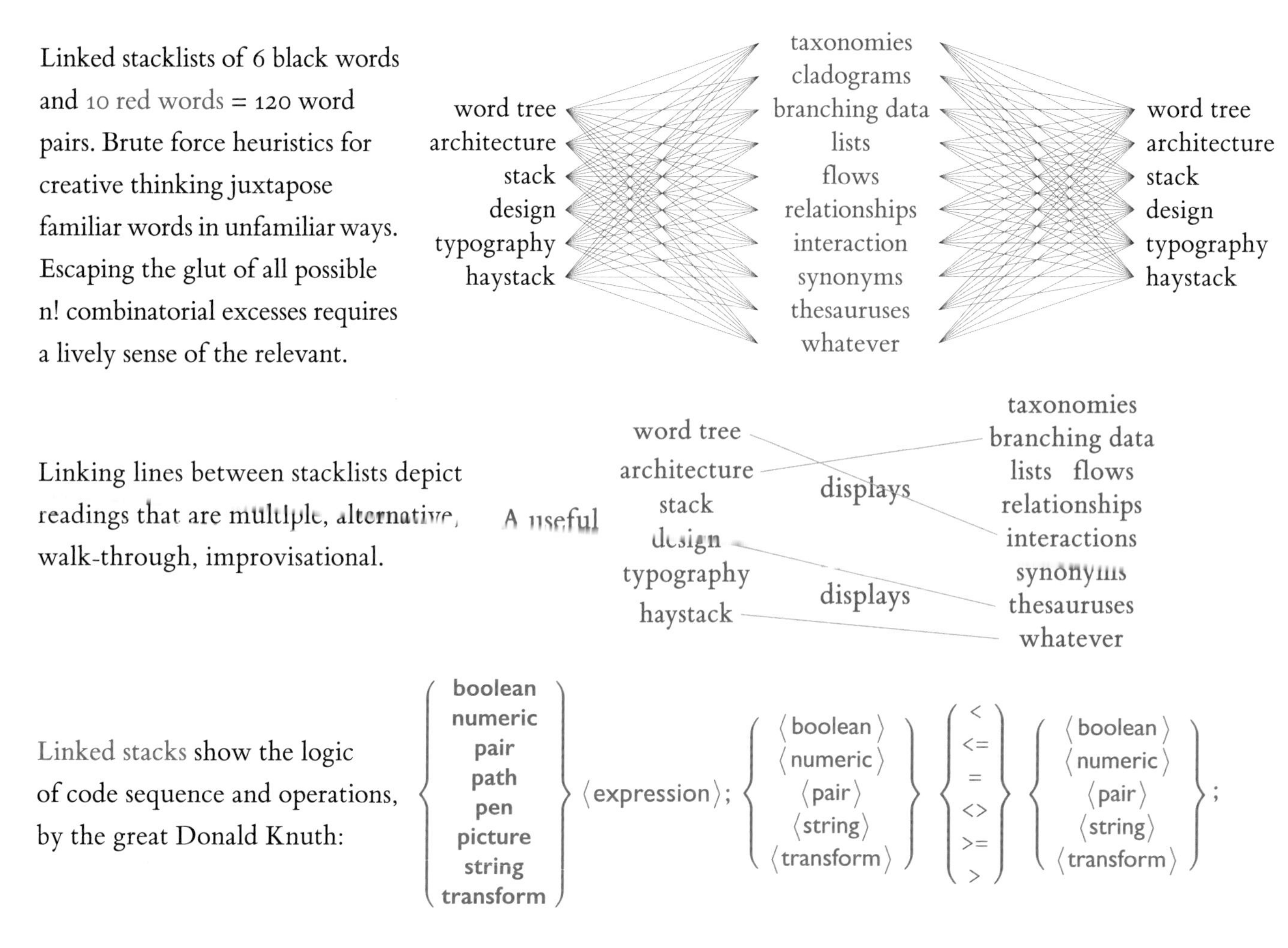

Upon discovering 100s of fabricated research reports published by suspect scientific journals, Elisabeth Bik sketched out a 7 stacklist Fake Research Title Generator (like other generators creating a polyphony of syntactically correct nonsense sentences constructed by walks through stacklists):

insert name of molecule:	**pick a verb:**	**choose 1 or 2 cellular processes:**	**choose cancer or cell type:**	**pick your connector word:**	**choose a verb (present participle):**	**insert name of miRNA or pathway:**
	alleviates	apoptosis			activating	
	attenuates	autophagy	lung cancer	by	down-regulating	
<protein name>	governs	inflammation	incidentaloma	by means of	inhibiting	<miRNA name>
<drug name>	inhibits	invasion	medulloblastoma	through	interfering with	<pathway name>
<RNA name>	prevents	migration	renal carcinoma	via	modulating	<protein name>
	relieves	proferation	ovarian cancer		targeting	
	suppresses	viability			up-regulating	

architectures
stacklists
STACKLIST PLAYGROUNDS DISPLAYING
haystacks
POETRY

lists
synonyms
taxonomies
flows
thesauruses

cloudy

WHATEVER

SHORT INLINE STACKLISTS SHOW STATISTICAL CONFIDENCE LIMITS IN TABLES AND SENTENCES

Component	A Stars	F Stars	Early G Stars
baryons	$0.088^{+0.006}_{-0.006}$	$0.088^{+0.007}_{-0.007}$	$0.085^{+0.007}_{-0.006}$
DM	$0.038^{+0.012}_{-0.015}$	$0.019^{+0.012}_{-0.011}$	$0.004^{+0.01}_{-0.004}$

TABLE II. Posteriors on the total baryonic and DM halo density at the midplane, in units of $M_{\odot}/pc^3$.

Lauren Anderson, et al, "Clustering of galaxies in the Baryon Oscillation Spectroscopic Survey," arXiv:2014

We find that the multipole results are slightly more precise, on average. We obtain tight constraints on both $\alpha_{||}$ and $\alpha_{\perp}$. In particular, $\widetilde{\Delta\sigma_{\alpha_{||}}} = -0.008^{-0.008}_{+0.007}$ and $\widetilde{\Delta\sigma_{\alpha_{\perp}}} = -0.003^{+0.003}_{-0.003}$ pre-reconstruction and post reconstruction, while the median difference in best-fit values are $\widetilde{\Delta\alpha_{||}} = -0.005^{+0.005}_{-0.004}$ and $\widetilde{\Delta\alpha_{\perp}} = -0.002^{+0.001}_{-0.002}$. As measurements from the two fitting methodologies are clearly correlated, it is not surprising that the obtained precisions on the αs are similar.

Katelin Schutz, Tongyan Lin, Benjamin R. Safdi, Chih-Liang Wu, "Constraining a Thin Dark Matter Disk with Gaia," arXiv: 2012

HORIZONTAL INLINE STACKLISTS WITH CONTENT SLASHES/LINEBREAKS

Eye-brain systems routinely detect and reason about
edges margins perimeters contours outlines fences boundaries
transitions changes differences convergences divergences collisions.

At the borderline and the borderspace between
meaning and space presence and absence figure and ground this and that us and them
where verbs/actions/interactions/conflicts all thrive and eyes see/think/design/produce.

STACK OF 13 SPACED-OUT DATA-SENTENCES NARRATE THE SEQUENCE OF SPORTS EVENTS AND HOW THE SCORE ACCUMULATED

Scoring in a baseball game is tracked by a stack of 13 sentences and 44 numbers. Each sentence describes who did what when, followed by a full line-space. The running score (at far right) accumulates to the final result.

Viewed by millions, similar templates narrate all sorts of worldwide sports events, often as they occur in real-time.

Scoring summary, Chicago Cubs vs. Cleveland Indians, game 7

			C Chicago Cubs (sans)	C Cleveland Indians (serif)
1st	CHICAGO	Fowler homered to center (410 feet)	1	0
3rd	CLEVELAND	Santana singled to right, Crisp scored	1	1
4th	CHICAGO	Russell hit sacrifice fly to center, Bryant scored, Zobrist to second	2	1
4th	CHICAGO	Contreras doubled to deep center, Zobrist scored	3	1
5th	CHICAGO	Báez homered to right center (408 feet)	4	1
5th	CHICAGO	Rizzo singled, Bryant scored, Rizzo to second advancing on throw	5	1
5th	CLEVELAND	Kipnis and Santana scored on Lester's wild pitch	5	3
6th	CHICAGO	Ross homered to center (406 feet)	6	3
8th	CLEVELAND	Guyer doubled to right center, Ramírez scored	6	4
8th	CLEVELAND	Davis homered to left (364 feet), Guyer scored	6	6
10th	CHICAGO	Zobrist doubled to left, Almora Jr. scored, Rizzo to third	7	6
10th	CHICAGO	Montero singled, Rizzo scored, Zobrist to third, Russell to second	8	6
10th	CLEVELAND	Davis singled to center, Guyer scored	8	7 final

POLYPHONY IN INTERLINEAR SENTENCES IN A SERIES OF SHORT STACKS

Glossing and parsing of the Egyptian *Book of the Dead*
show interlinear transcriptions, transliterations, translations.

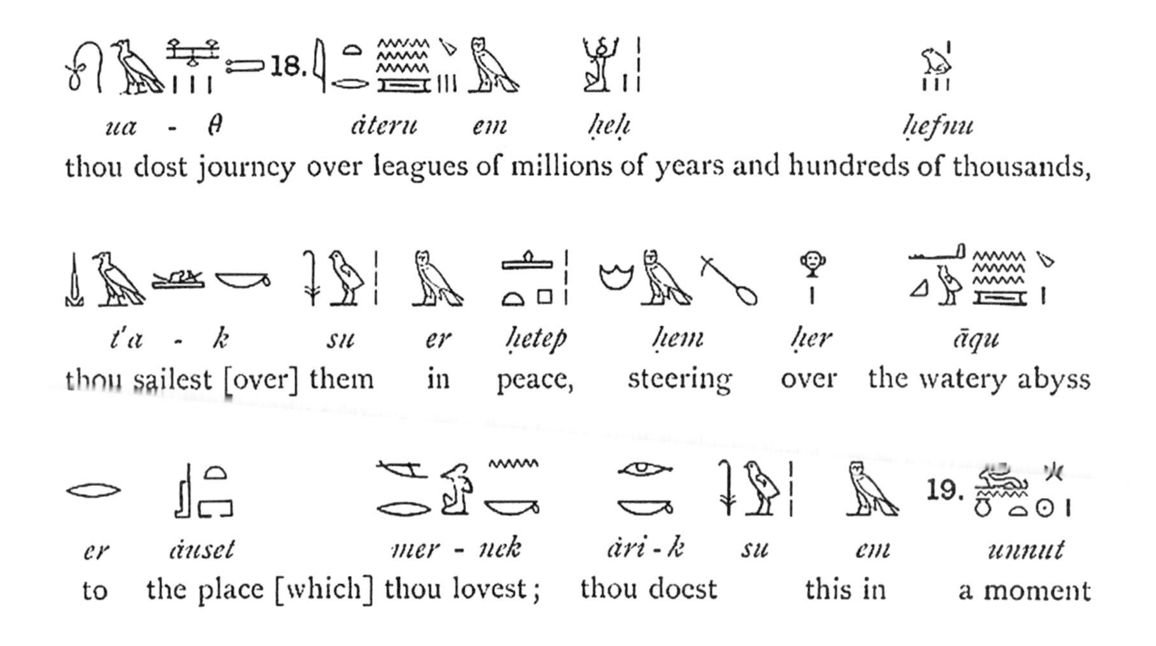

18.
ua - θ āteru em ḥeḥ ḥefnu
thou dost journey over leagues of millions of years and hundreds of thousands,
t'a - k su er ḥetep ḥem ḥer āqu
thou sailest [over] them in peace, steering over the watery abyss
19.
er āuset mer - nek āri - k su em unnut
to the place [which] thou lovest; thou doest this in a moment

GRAPHICAL EQUATIONS EMBEDDED IN GRAPHICAL SENTENCES

33 short stacklists and 13 short inline-lists describe how subatomic particles behave.
Scientists do their typesetting/equations/diagrams in open source LaTeX and extentions
(and MIT undergraduates do their problem sets in LaTeX).

$$-\Gamma_1^{(l)} = \frac{1}{l-1}\left\{ \frac{\delta - \Gamma_1^{(l-1)}}{\delta\, 1 \leftarrow 2} - \frac{\delta - \Gamma_1^{(l-1)}}{\delta\, 1 \sim 2} + \frac{1}{2} \frac{\delta^2 - \Gamma_1^{(l-1)}}{\delta\, 1 \leftarrow 2\ \delta\, 3 \leftarrow 4} + \sum_{k=2}^{l-2}\left[\frac{\delta - \Gamma_1^{(k)}}{\delta\, 1 \sim 2} \frac{\delta^2 - \Gamma_1^{(l-k)}}{\delta\, 3 \leftarrow 4\ \delta\, 5 \leftarrow 6} + 2\, \frac{\delta - \Gamma_1^{(k)}}{\delta\, 1 \sim 2} \frac{\delta - \Gamma_1^{(l-k)}}{\delta\, 3 \leftarrow 4} \right]\right\}$$

$$\frac{\delta - \Gamma_1^{(3)}}{\delta\, 1 \leftarrow 2} = - \ldots - \ldots + \ldots - \ldots ,$$

$$\frac{\delta^2 - \Gamma_1^{(3)}}{\delta\, 1 \leftarrow 2\ \delta\, 3 \leftarrow 4} = - \ldots - \ldots - \ldots + \ldots + \ldots + \ldots + \ldots + \ldots + \ldots - \ldots - \ldots - \ldots$$

"Functional Closure of Schwinger-Dyson Equations in Quantum Electrodynamics: Generation of Connected and One-Particle Irreducible Feynman Diagrams," Axel Pelster, Hagen Kleinert, Michael Bachman

HAND-DRAWN CONTENT-RESPONSIVE GRIDS BREAK CONVENTIONAL GRIDS

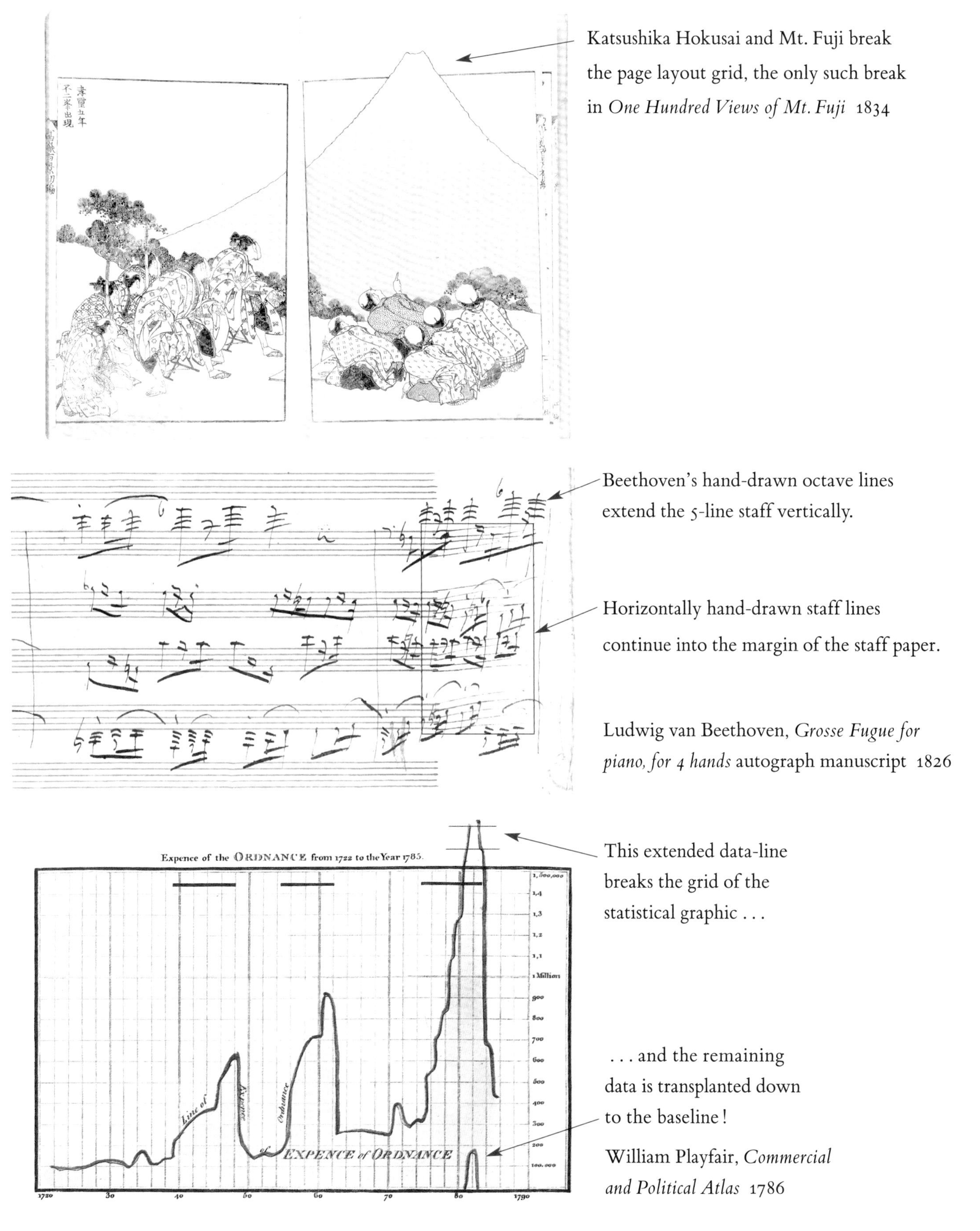

Katsushika Hokusai and Mt. Fuji break the page layout grid, the only such break in *One Hundred Views of Mt. Fuji* 1834

Beethoven's hand-drawn octave lines extend the 5-line staff vertically.

Horizontally hand-drawn staff lines continue into the margin of the staff paper.

Ludwig van Beethoven, *Grosse Fugue for piano, for 4 hands* autograph manuscript 1826

This extended data-line breaks the grid of the statistical graphic . . .

. . . and the remaining data is transplanted down to the baseline!

William Playfair, *Commercial and Political Atlas* 1786

ON FOXES AND HEDGEHOGS BY ISAIAH BERLIN

Isaiah Berlin, *The Hedgehog and the Fox: An Essay on Tolstoy's View of History,* 1953

> There is a line among the fragments of the Greek poet Archilochus which says: 'The fox knows many things, but the hedgehog knows one big thing.'[1] Scholars have differed about the correct interpretation of these dark words, which may mean no more than that the fox, for all his cunning, is defeated by the hedgehog's one defence. But, taken figuratively, the words can be made to yield a sense in which they mark one of the deepest differences which divide writers and thinkers, and, it may be, human beings in general. For there exists a great chasm between those, on one side, who relate everything to a single central vision, one system, less or more coherent or articulate, in terms of which they understand, think and feel – a single, universal, organising principle in terms of which alone all that they are and say has significance – and, on the other side, those who pursue many ends, often unrelated and even contradictory, connected, if at all, only in some *de facto* way, for some psychological or physiological cause, related by no moral or aesthetic principle. These last lead lives, perform acts and entertain ideas that are centrifugal rather than centripetal; their thought is scattered or diffused, moving on many levels, seizing upon the essence of a vast variety of experiences and objects for what they are in themselves, without, consciously or unconsciously, seeking to fit them into, or exclude them from, any one unchanging, all-embracing, sometimes self-contradictory and incomplete, at times fanatical, unitary inner vision. The first kind of intellectual and artistic personality belongs to the hedgehogs, the second to the foxes; and without insisting on a rigid classification, we may, without too much fear of contradiction, say that, in this sense, Dante belongs to the first category, Shakespeare to the second; Plato, Lucretius, Pascal, Hegel, Dostoevsky, Nietzsche, Ibsen, Proust are, in varying degrees hedgehogs; Herodotus, Aristotle, Montaigne, Erasmus, Molière, Goethe, Pushkin, Balzac, Joyce are foxes.

Few contrasts in cognitive style are so knowing and known as foxes vs. hedgehogs. Isaiah Berlin's text is a bit of a muddle upon first and second readings. The parallel structure begins *fox first, hedgehog second:* "The fox knows many things, but the hedgehog knows one big thing." After another *fox-first,* the order is reversed to 4 *hedgehog-firsts.* A stacklist shows the complete sequence –

broken parallelism	*fox hedgehog* *fox hedgehog* *hedgehog fox* *hedgehog fox* *hedgehog fox* *hedgehog fox*	resulting in unnecessary confusion.

FOX AND HEDGEHOG STACKS CLARIFY STRUCTURE AND MEANING

'The **fox** knows many things,	but the **hedgehog** knows one big thing.'
The **fox**, for all its cunning,	is defeated by the **hedgehog's** one defense.
Foxes who pursue many ends, often unrelated and even contradictory, connected, if at all, only in some *de facto* way, for some psychological or physiological cause, related by no moral or aesthetic principle.	
Foxes lead lives, perform acts and entertain ideas that spin outward rather than inward; their thought is scattered or diffused, moving on many levels, seizing upon the essence of a vast variety of experiences and objects for what they are in themselves, without, consciously or unconsciously, seeking to fit them into, or exclude them from, any one unchanging, all-embracing, sometimes self-contradictory and incomplete, at times fanatical, unitary inner vision.	**Hedgehogs** who relate everything to a single central vision, one system, less or more coherent or articulate, in terms of which they understand, think and feel – a single, universal, organising principle in terms of which alone all that they are and say has significance.
foxes	**hedgehogs**
Shakespeare Herodotus Aristotle Montaigne Erasmus Molière Goethe Pushkin Balzac Joyce	Dante Plato Lucretius Pascal Hegel Dostoevsky Nietzsche Ibsen Proust

TWO STACKLIST SENTENCES EVOKE A THEORY OF KNOWLEDGE

How do you / I / they know that?   How could you / I / they possibly know that?

To reason about evidence and conclusions, to evoke self-awareness about the truth of our knowledge, to measure explanatory depth, ask *How do I know that? How do we know what we don't know?* Feynman's law: 'The first principle is that you must not fool yourself, and you are the easiest person to fool.' Ask others *How do you know that?*

Then, thought experiments about knowledge: *How could anyone possibly know that?* What research design would produce credible evidence for the claimed knowledge? If none, the claimed knowledge is not even wrong – for it is impossible to prove/disprove.

GRAPHICAL SENTENCES
CONNECTING CONVERSATION

Sentences flow from speaker to listener in this imagined meeting in ~1315 of Catalan author and illustrator Ramon Llull, his editor Thomas Le Myésier, and the Queen of France with her colleagues. Queen Jeanne d'Évreux receives manuscripts pitched by author and editor. To depict this operatic fantasy, words and images are closely combined, as flowing words connect 2 speakers with a listener. Such integration is likely when words and images are produced by the same hand – in this case, the illustrator's hand. At right, the original Latin version:

Thomas preserved the main points, did not
change the text. Done very well indeed!
He made it easier for others.
Your kind majesty, receive this small gift,
which is like a little friend.
manuscript
abstract
abridged
manuscript
first step of project, excerpts
from Llull's books, collected
and organized following plan

TWO VERY GOOD GRAPHICAL SENTENCES

Maira Kalman arranged 2 graphical sentences with 3 stacklists and 3 or more perspective viewpoints. Location, size, and content of words gather together meaning.

Buster says, "Nuts to Pete."
Now
Bennie
has no
money (NONE) to buy Buster
a new ball which, you will remember, Pete
ate
many
letters
ago.
n
N
n

Words narrow in these stacks, reading upward and moving deeper into the scene.

Beginning a long sentence, a skewed stack reads down and sideways.

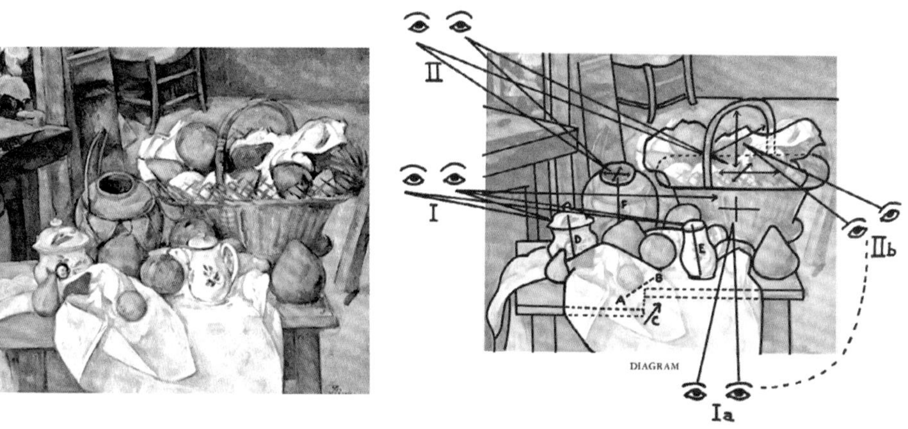

Remodeling Cubism (which remodeled painting), Erle Loran and John Berger suggest early Cubism consists of *explanatory diagrams* describing the history of the artist's views of a painted scene! "The metaphorical model of Cubism is the *diagram*, a visible, symbolic representation of invisible processes, forces, structures."

How do you / I / they know that? How could you / I / they possibly know that?

To reason about evidence and conclusions, to evoke self-awareness about the truth of our knowledge, to measure explanatory depth, ask *How do I know that? How do we know what we don't know?* Feynman's law: 'The first principle is that you must not fool yourself, and you are the easiest person to fool.' Ask others *How do you know that?* Then, thought experiments about knowledge: *How could anyone possibly know that? What research design would produce credible evidence for the claimed knowledge?* If none, the claimed knowledge is not even wrong – for it is impossible to prove or disprove.

The fundamental principles of analytical thinking: reason about causality and mechanism, explain something, make comparisons, assess credibility of measurements and inferences, validate findings, enforce integrity and honesty. These *principles cannot be altered or repealed* by assumptions, by any discipline or specialty, political and intellectual fashions, marketing or monetizing, what sponsors desire or demand, by anything you do, think, believe, hope.

WHY RESEARCH ON HUMANS IS WAY MORE DIFFICULT THAN ROCKET SCIENCE

Nature's mathematical laws apply to every particle everywhere forever. From tiny particles and undetectable waves up to the entire universe, all measurements testify about those laws. Truth and exactitude are always present – and so real science is easy, compared to understanding biological systems and human behavior, which lack mathematical laws.

In biological systems, DNA masterplans are a cumulative tangle of local random evolutionary hacks, work-arounds, mutations. Unlike universal physical laws, 'living organisms are historical structures, literally creations of history. They represent a patchwork of odd sets pieced together when and where opportunities rose.' Some biological complexities appear unfathomable and irredeemable: 'random replicative mutations in stem cell divisions (bad luck) are largely responsible for variations in cancer risk.'

Research on humans involves the space-time mash-ups of biological systems – and humans who act, plan, think, connive, cheat, think about thinking, and fail to recognize their ignorance. Leo Tolstoy's *War and Peace* described the difficulties of making causal inferences about human behavior:

> 'When we say that Napoleon *commanded* armies to go to war, we combine in one simultaneous expression a whole series of consecutive commands dependent on one another. Our false idea that an event is caused by commands that precede it, out of 1000s of commands those few that were executed and consistent with the event – and we forget about the others that were not executed because they could not be.'

With greater understanding of human/biological complexities, life expectancy has *doubled globally since 1900.* In 1950, worldwide life expectancy was 48 years, and in 2019, 71 years. Advances in public health, science, economic development, education, and evidence-based medicine produced these unprecedented gains in human history. Then came the pandemic.

4 DATA ANALYSIS WHEN THE TRUTH MATTERS: ON THE RELATIONSHIP BETWEEN EVIDENCE AND CONCLUSIONS.

TAKING ANONYMOUS STATISTICAL LIVES AS SERIOUSLY AS IDENTIFIED INDIVIDUAL LIVES

There is a common preference to rescue and extend *named individual lives*, no matter what the cost. Yet comparable investments might save millions of *anonymous invisible statistical lives*, since the cost extending a statistical life is often small compared to extending a named life.

named and nameless lives visable and invisable lives private vs. public interests insiders and outsiders short vs. long time-horizons rescue treatments vs. prevention personalized precision medicine (n = 1) vs. vaccination (n = 3,000,000,000) local optimizing vs. global pessimizing proprietary vs. open source hedgehogs and foxes

A mistake in the operating room can threaten the life of one patient, a mistake in statistical analysis or interpretation can lead to hundreds of early deaths. So it is odd that, while we allow a doctor to conduct surgery only after years of training, we give software packages – SPSS [Python, R, MATLAB, Machine Learning, et al] – for statistical analysis to almost anyone. ANDREW VICKERS

THE STANDARDS OF STATISTICAL REASONING ABOUT THE TRUTH ARE UNIVERSAL

Increasing knowledge begets increasing specialization and narrower scope of understanding. Statistics, as the practice of planning experiments and observations and of interpreting data, has a common relation to all sciences. Unification will be more easily attained if the logical framework of the individual sciences can be identified and isolated from their factual content. FRANCIS ANSCOMBE, ET AL.

A statistician I knew once replied to a brain surgeon whose hobby was statistics that hers was brain surgery. STEPHEN JOHN SENN

Although we often hear that data speak for themselves, their voices can be soft and sly. It is easy to lie with statistics; it is easier to lie without them. FREDERICK MOSTELLER

Confirmation bias is the tendency to search for, interpret, favor, and recall information so as to confirm one's pre-existing beliefs or hypotheses. WIKIPEDIA *It is a principle that shines impartially on the just and the unjust alike that once you have a point of view all history will back you up.* VAN WYCK BROOKS

Bias can creep into the scientific enterprise in all sorts of ways. But financial conflicts are detectable definitively and represent a uniquely perverse influence on the search for scientific truth. COLIN BEGG

It is simply no longer possible to believe much of the clinical research that is published, or to rely on the judgment of trusted physicians or authoritative medical guidelines. I reached this conclusion slowly and reluctantly during my 2 decades as an Editor of The New England Journal of Medicine as drug companies asserted more power, and began to treat researchers as hired hands. MARCIA ANGELL

REMODELING DATA MEASUREMENT AND ANALYSIS: TAKING STATISTICAL LIVES AS SERIOUSLY AS INDIVIDUAL LIVES.

Cardiac Surgery: Safeguards and Pitfalls in Operative Technique compiled by Siavosh Khonsari and Colleen Sintek is an intense encyclopedia of 3,000 alerts and warnings: ⊘ = *avoid this grave error, how to prevent it, and if it occurs, what to do,* and the less severe alerts ! = *warning,* NB = *note well.* Here is one alert from 3,000:

A needle inadvertently picking up the bottom of an artery wall

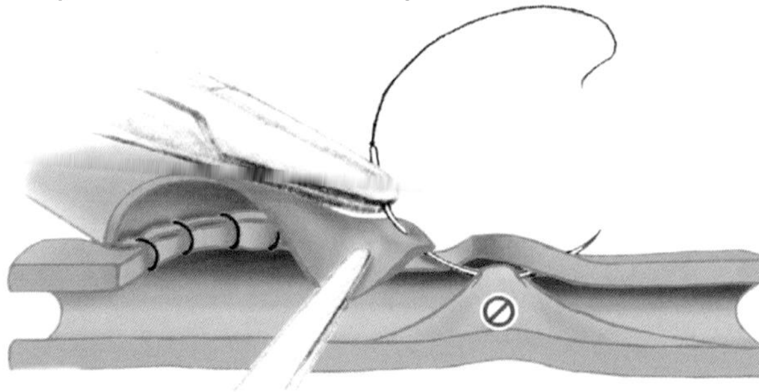

⊘ INADVERTENT SUTURE OF POSTERIOR WALL

"The anastomosis toe is the most critical part because it determines graft outflow capacity. When the artery lumen is small or visibility is suboptimal, the needle may pick up the artery posterior wall. An appropriately sized plastic probe passed for a short distance into the distal artery allows precise placement of sutures and prevents occurrence of this complication."

Imagine now another encyclopedia, *Data Measurement and Analysis: 1,000 Safeguards, Pitfalls, and Cheats in Statistical Practice,* where quality control for statistical lives would seek to match quality control for named lives in cardiac surgery. An encyclopedia entry:

⊘ INAPPROPRIATE IMAGE DUPLICATION

In research papers, fraud is often detected by carefully examining images and data graphics. In this retracted paper, apparently falsified Western Blot Tests were created by copying nearby blots, and then disguised by flipping and mirror-reversing. This problem is obvious because the blot *labels* were inadvertently flipped and reversed, resembling upside-down Russian words!

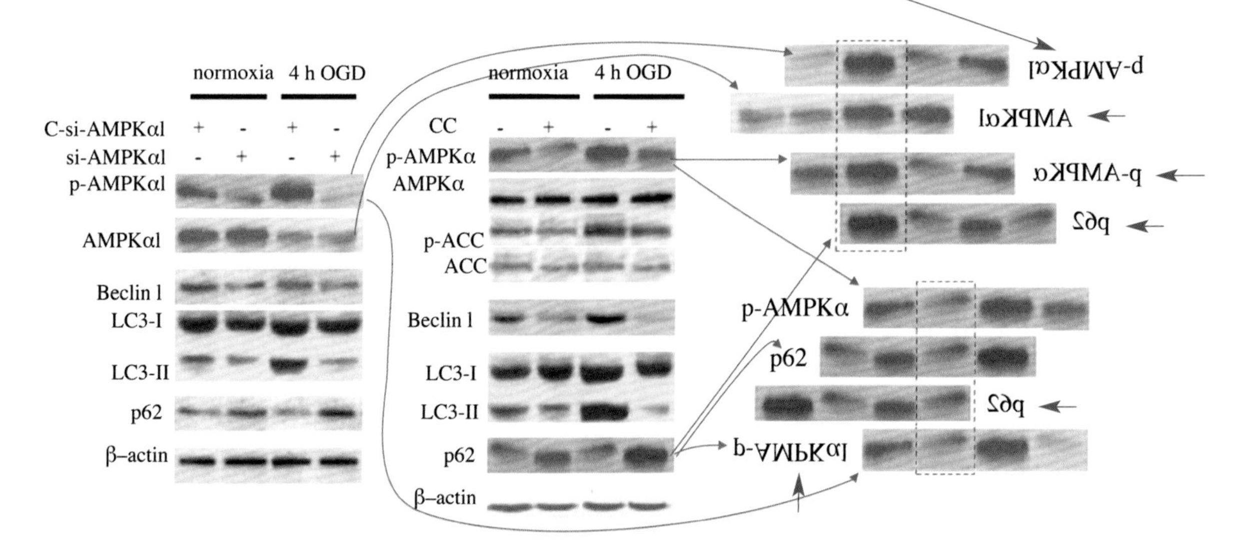

PREVALENCE OF MEASUREMENT AND DATA ANALYSIS PITFALLS IN MEDICAL RESEARCH

Examining 1000s of published studies, meta-researchers measure prevalences of pitfalls and cheats in data collection/analysis. The prevalence rates are appalling. (Should we be thankful that 95% of published medical research vanishes, unread and uncited even by the authors' mothers?) For example, early reports of medical interventions make enthusiastic claims that never again will be achieved. As evidence improves, harms may eventually exceed benefits (eg, baby aspirin for heart attack prevention) – and prevalence of regression toward the truth is perhaps 80%. Here are prevalence rates of major problems in medical research: randomization failures, inability to undo compromised research in a timely way, image integrity, measurement quality – more entries for a proposed encyclopedia *Data Measurement and Analysis: 1,000 Safeguards, Pitfalls, and Cheats in Statistical Practice.*

⊘ COMPROMISED RANDOMIZING: THE MEDITERRANEAN DIET STUDY AND ITS 267 OFFSPRING
#☠☆¡!! RETRACTION SYNDROME ☆ INDEPENDENT AUDITS OF RESEARCH = A NECESSITY

'Precise fulfillment of randomization protocols assigning subjects to treatment vs. control is critical to research design. Imprecise randomization causes serious complications,' reports the thorough John Carlisle, who detected incorrect randomizing in the famous Mediterranean Diet study (NEJM 2013). This was retracted, corrected data analyzed and published – with somewhat weaker evidence for Mediterranean diets (NEJM 2018). 267 secondary articles based on original incorrect data still remain at large. Prevalence of randomizing mess-ups are based on 5,087 articles; note strongly-worded titles concerned about credibility and truth:

"Data fabrication and other reasons for non-random sampling in 5,087 randomized, controlled trials in anesthetic and general medical journals," John B. Carlisle, *Anesthesia* 72 (2017), 931-935. "PREDIMED trial of Mediterranean diet: retracted, republished, still trusted?" Arnav Agarwal and John P.A. Ioannidis, *BMJ*, 7 Feb 2019.

⊘ INAPPROPRIATE IMAGE DUPLICATION: 3.8% AND 6.1% PREVALENCE RATES

"Images from 20,621 papers published in 40 scientific journals 1995-2014 were visually screened: 3.8% of published papers contained problematic images, half exhibited features suggesting deliberate manipulation." Another study reviewed 960 papers from *Molecular and Cellular Biology* 2009-2016 and found 6.1% (59 of 960) "papers to contain inappropriately duplicated images, leading to 41 corrections, 5 retractions."

"Prevalence of inappropriate image duplication in biomedical research publications," Elisabeth Bik, Arturo Casadevall, Ferric Fang, *ASM mBio7*, 2016; "Analysis and correction of inappropriate image duplication," Elisabeth Bik, et al *MCB* (2018).

⊘ GENE NAME CONVERSION ERRORS IN EXCEL: 704 ARTICLES IN 18 GENOMICS JOURNALS

In default settings, Microsoft Excel converted gene names with 3 letters and 2 numbers to dates and floating-point numbers. "A programmatic scan of leading genomics journals reveals 20% of papers with supplementary Excel gene lists contain erroneous gene name conversions." This error was known to insiders in ~2004; alas from 2005-2015, 704 papers published in 18 journals were affected.

"Gene name errors widespread in scientific literature," Mark Ziemann, Yotam Eren, Assam El-Osta, *Genome Biology* 17 (2016). Update: James Vincent, "Scientists rename human genes to stop Microsoft Excel from misreading them as data: Sometimes it's easier to rewrite genetics than update Excel," *The Verge*, August 6, 2020.

FACING UP TO MEASUREMENTS

☆☆☆ OBSERVE DATA COLLECTION AT THE MOMENT OF MEASUREMENT

See, observe, learn how data are collected at moment of measurement. "You never learn more about a process than when you directly observe how data are actually measured," said Cuthbert Daniel, a superb applied statistician. See with fresh eyes. Walk around what you want to learn about. Talk to those who do measurements. See how numbers came to be. Are those measuring skilled/honest/biased/incompetent/tired and emotional/sloppy? Are errors and artifacts in measurements assessed? How are outliers adjudicated? Those directly observing medical measurements may well rightly conclude that medical care/research is a two-digit science on a very good day. On other days, getting the sign right is an achievement.

Example: big company polluted big river, environmental agencies forced polluters to clean up and monitor progress by daily water samples. Small boat goes out to take sample, dipping into the water *after looking around for the cleanest water.* Statisticians call this 'sampling to please.' Observing this data collection reveals the early limits of self monitoring, and that people can't keep their own score.

Read service manuals on measurement practices and their artifacts. Do detective work. Nurses and techs make measurements all day long; ask about quality, errors, relevance, false alarms, and duplicate/unnecessary/lucrative measurements that signal over-diagnosis. People love to talk about their work, let them do the talking. Ask others the same questions. Check, verify, recheck reports. From 34 hours (2019-2021) of my interviews with ICU/oncology nurse (U.S.):

Q. End-of-life care in practice? *A. "Oncologist pulled near-death patient out of hospice who reportedly 'looked better.' What did patient do, smile – vitals are not even measured in hospice! Patient sent to oncology floor, I refused to administer chemo because pt was about to die, oncologist backed down when I threatened to inform Patient Ethics Committee, pt died that night."* Q. With your additional responsibilities as floor nurse-in-charge, how much more you paid? *"$1.00 per hour at both hospitals where I've worked."* Q. How can that be ?! *"94% of nurses are females"* Q. Doctor quality? *"Big range, a few surgeons are awful, everyone knows who, some anesthesiologists won't work with them."* Q. Cause of Death forms? *[long discussion = death adjudication is uncertain]* Q. (On covid, Spring, 2021 interviews): *"Nurses the only people seeing patients. Proning, placing pt stomach down, reduced ventilator use. Hospital supplied one N95 mask for 2 weeks reuse. PPE shortage. No hazard pay, they said medical center 'had lost $100s millions.' 17 nurses on my floor got covid-19."* Q. Biggest thing you learned? *"[Hospital Company] does not care about nurses."* Q. What's new? *"For Nurses week, Medical Center gave Nurses a rock in paper bag, note saying 'make a Gratitude Rock, paint this, give to a friend, express your gratitude.' "*

☆☆☆ OBSERVE DATA AT MOMENT OF MEASUREMENT **NOT AT MOMENT OF PUBLICATION**

Getty Research Institute, Los Angeles (2014.R.20). Photograph: Shunk-Kender © J. Paul Getty Trust

☆☆☆ TRACK THE ORIGINS AND LIFE HISTORIES OF MEASUREMENTS

Why is the measurement made? Who looks at the measurements, when, where, why? What are consequences, harms, benefits? Where did the data go, what's it doing now? For medical measurements, track the money: who profits how much? Behind the scenes, engineering/technical manuals frankly describe measurement complexities and problems.

☆☆☆ ACTUAL MEASUREMENTS VS. PUBLISHED DATA: INVESTIGATION OF SCIENTIFIC MISCONDUCT

FROM "THE REPORT OF THE INVESTIGATION COMMITTEE ON THE POSSIBILITY OF SCIENTIFIC MISCONDUCT IN THE WORK OF HENDRIK SCHÖN AND COAUTHORS, APPENDIX E: ELABORATED FINAL LIST OF ALLEGATIONS." LUCENT TECHNOLOGIES, SEPTEMBER 2002. EDITED.

"Can the data presented be traced back to primary data, free of any data processing or other manipulation?

It is a well-established tenet of science that clear records should be kept. Only credible, primary data can provide unambiguous corroborating evidence for published data. An understanding of the procedures of data acquisition and analysis also provides a context within which possibly mitigating circumstances can be assessed. It is worth emphasizing that the retention of primary data, together with adequate record keeping, are necessary to the ordinary conduct of science, not simply for the examination of possible wrongdoing. In the conduct of research, new questions arise that require a revision of the original analysis, and thus require a return to the primary data. Failure to keep primary data and records for a reasonable time is, by itself, a threat to the health of the scientific enterprise. This remains as true in the computer age as it has been in the past.

Is there clear evidence that the data do not come from the measurements described?

This evidence takes different forms: *Data Substitution*, in which data sets for distinct experimental conditions show unreasonable similarity to each other, in some cases after multiplying one data set by a constant factor; *Unreasonable Precision*, in which a data set agrees better with a simple analytic expression than would be expected from the measurement accuracy; and *Contradictory Physics*, in which the data appear to be inconsistent with prevailing scientific understanding and description of the measurement. Many great discoveries in science would at first have been included in the Contradictory Physics category, so the Committee has set aside all but a few especially problematic examples. But extraordinary results demand extraordinary proof. Unless special diligence is demonstrated, results that contradict known physics suggest simple error, self-deception, or misrepresentation of data.

If the data are not valid, are there mitigating circumstances that explain how the data came to be misrepresented?

For example, a clerical error in including the wrong data in a figure represents poor procedures, but not misconduct. But such innocent explanations may require understanding of the state of mind of the authors at the time the data were prepared, and this cannot be determined definitively. It must be noted that the credibility of a particular innocent explanation depends on the overall credibility of the scientist in question. This in turn depends on whether there is an unreasonable number of problems or a pattern of questionable practices. Rather, the problems with the data are already established, and the question is whether many improbable, innocent explanations should be accepted."

☆☆☆ 'YOU CAN OBSERVE A LOT BY JUST WATCHING. OF COURSE YOU DO HAVE TO KNOW WHAT TO WATCH, AND YOU DO HAVE TO KEEP WATCHING'

Cuthbert Daniel: "The name 'statistician' meant something different to management from what it meant to me. For them, it was complete records of what the plant was doing at all points, all times: a graph of U235 concentration versus Building Numbers from 1 to 47 was a curve that went from 0.7% (where it starts) up to >90%. One of my jobs was to keep weekly records of average measurements made by mass spectroscopy in each building. After a month I noted while the graph went steadily up, one building showed flatness, not a step. Instead of rushing to management saying something is wrong in building 41, I went to the smartest Process Engineer and said, 'If a building wasn't working at all, how would it show?' He said, "That's easy. Valve F43 is open, bypassing the barriers.' Then I was alarmed, went to the plant manager, said 'Valve F43 has been open for the last month.' That's the only time I ever got any attention from him and I got it then. How did I know that? *It was a graph, no exact numbers, no math.* Only weekly building averages. This exemplifies the Yogi Berra Principle: *'You can observe a lot by just watching.' Of course you do have to know what to watch, and you do have to keep watching!"* ET, "Conversation with Cuthbert Daniel," *Statistical Science*, 1988, 413-424

⚠ ARE MEASUREMENTS REAL, OR JUST BANG-BANG DUPLICATES?

An observer measures and records a number and then a few seconds later records another. These two measurements are not independent if the process changes hourly. Knowledge of the first number gives full knowledge of the second, and the number of independent measurements is not two but one. In econometrics, this is called *autocorrelation or serial correlation; in experimental design, bang-bang duplicates or pseudo-replications.* Duplicates do show up in mouse studies: if a researcher does an experiment on 3 mice and measures the same variable on each mouse 30 times, the sample size is not 90. The only way to achieve statistical significance with 3 mice is if one of them turns into a cat.

⚠ CONSEQUENCES OF A SPREADSHEET FILLED WITH NOISE

ALL SPORTS COMMENTARY
xkcd

Spreadsheets of random numbers contain no viable findings, except that you've got the wrong spreadsheet. Dr. Confirmation Bias and Dr. pHacker fabricate stories based on noise. WEIGHTED RANDOM NUMBER means that while some sports competitors are better than others, the exciting fine-grain local variations around averages are random, luck, coincidences, cheats, outliers, miracles. Weak evidence spawns big attitudes, a rage to conclude: "Ignorance more frequently begets confidence than does knowledge," said Charles Darwin.

⚠ YOUR DATABASE MAY NOT CONTAIN THE TRUTH OR THE ANSWERS TO ALL OF YOUR QUESTIONS

Is the database at hand capable of answering the research questions at hand? A database may not contain the relevant explanatory variables, a devastating constraint. And why should any data set contain all relevant known knowns and unknown knowns, and known unknowns, and the most difficult, unknown unknowns? Many data analyses are model specification searches – let's try this, let's try that. But *this* and *that* might not even be in the database. Get more/different data. Far better, have an independent explanatory theory, as in real science. Truth = explanatory theory, evidence, independent replication, and quality/honesty/integrity in conducting research. Enemies of the truth: (1) no explanatory theory, (2) no empirical truth available for the topic at hand, (3) lack of relevant data, (4) incompetence, gullibility, lack of self-skepticism, (5) Drs. Confirmation Bias and their conflicts of interest, (6) cheats, lies.

⚠ 97% OF 565 MEDICAL RESEARCH REPORTS FAILED TO DEAL WITH MEASUREMENT ERRORS

A 2018 analysis examined practices in reporting measurement errors in 12 major medical and epidemiology journals: *44% of research articles mentioned measurement errors, and only 7% of those investigated or corrected the errors.*

Timo B. Brakenhoff, Marian Mitroiu, Ruth H. Keogh, Karel G.M. Moons, Rolf Groenwold, Maarten van Smeden, "Measurement error is often neglected in medical literature," *Journal of Clinical Epidemiology,* 2018.

⚠ IN HEALTHCARE, FALSE ALARMS ARE THE MOST PREVALENT AND LUCRATIVE MEASUREMENT ERRORS

Screening tests produce many false alarms, terrifying millions of healthy people. False alarms cascade into more tests. Mass screenings are now regarded as dubious – because of false alarms, harms, and failure to reduce all-cause mortality. Early diagnosis leads to early cures, or that patients just get the bad news sooner. Since survival time = time from diagnosis to death, early diagnosis can create statistical illusions of improved survival times. And false alarms, if their falsity is not detected, lead to treatments of patients for a disease they don't have. This entire gray area shows a mix of over-diagnosed, cured, indolent, incidentals, subclinical, and harmless cancers (many older people die *with* cancer, not *of* cancer). Data on the Number Needed to Treat (the number of patients needed to treat for a single favorable outcome) indicates that from 2 to 1000 cancer patients are treated for each one that benefits. Often far more are harmed than benefitted.

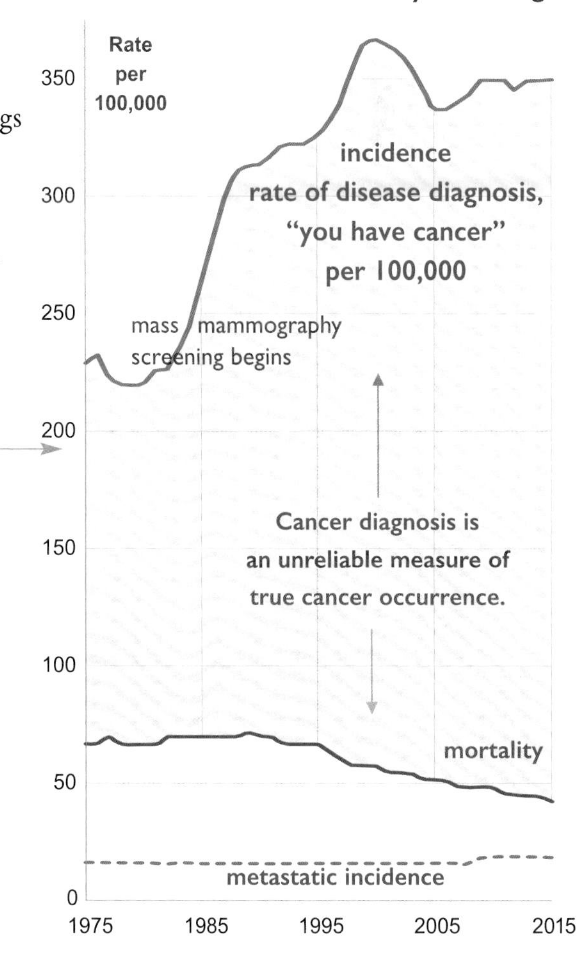

Data from 2014-2015 Surveillance Epidemiology End Results (SEER). Gilbert Welch, Barnett S. Kramer, William C. Black, "Epidemiologic Signatures in Cancer," *New England Journal of Medicine,* Oct 2019, redrawn. See also Andrea R. Marcadis, Jennifer L. Marti, Behfar Ehdaie, et al, "Characterizing Relative and Disease-Specific Survival in Early-Stage Cancers," *JAMA Internal Medicine,* December 9, 2019.

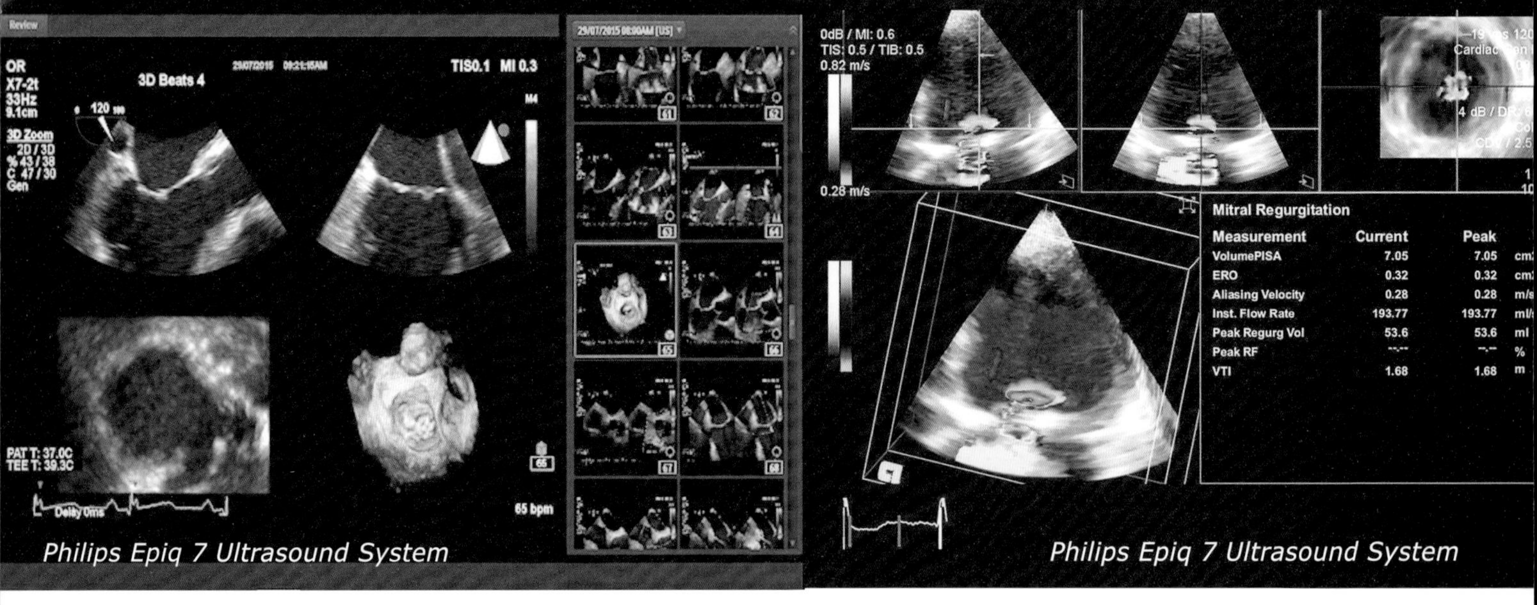

Philips Epiq 7 Ultrasound System

Philips Epiq 7 Ultrasound System

REAL-TIME DATA ANALYSIS WHEN THE TRUTH MATTERS: MULTIPLE 2D/3D MEASUREMENTS DESCRIBE PATIENT STATUS AND PERFORM LIVE QUALITY CONTROL DURING HEART SURGERY

On dividing a narrow space between the aorta and pulmonary artery:
Senior heart surgeon: "Divide exactly in the middle."
Student of heart surgery: "If I err, to which side should I err?"
Senior heart surgeon: "Don't err." DR. MARC GILLINOV

An excellent data visualization, transesophageal echocardiograph (TEE, pronounced "T-E-E") provides *live imaging and quantitative measurements during heart surgery, exactly at point of need.* Requiring substantial real-time signal-processing, TEE imaging is based on ultrasound data from a transducer/receiver placed within a few centimeters of the heart by going down the patient's esophagus, bouncing ultrasound waves off the heart, then converted to images and data.

When the heart surgery is almost completed, but the patient is still opened up, the surgical results up to this point are evaluated: the patient is taken off the heart-lung machine, the heart restarted, then TEE provides visual assessments on how the newly repaired heart is working. If an issue appears, the patient goes back on-pump, heart stopped again, issue repaired, patient then closed up. TEE images are displayed on a low-glare matte screen for surgeon and sonographer, and shown on two large glary display screens viewed by the 12 people in the OR. This is a superb visualization and data model: it stays close to the data, processes and analyzes data in real-time, leads to quality control and interventions, works hand-in-hand with the surgical team. (In some other medical situations, TEE is overused with few diagnostic benefits.)

EXCELLENT ASSESSMENT OF ARTIFACTS AND ERRORS IN 2D/3D TEE DATA

TEE technical manuals describe how 2D and 3D imaging models produce artifacts in daily use. Sonographers can identify image artifacts and then make corrections by adjusting the position of the ultrasound transducer/receiver. Note the attention to detail. *This is what serious empirical analysis of measurement errors looks like, and it is enormously better than the models of "error" in classical statistics.*

Look over next page with care, just read around the technical jargon.

"Acoustic Artifacts in 2D Imaging Acquisition

The transducer adds noise to the echo signal by beam-width effects, axial resolution limitations, frequency characteristics. Control choices made by sonographers affecting amplification, signal processing, and echo signal can lead to significant differences in echo data. **Acoustic saturation** occurs when received signals reach a system's high-amplitude limit, when the system is unable to distinguish or display signal intensities. At saturation, increased input will not increase output. **Aliasing** occurs when the detected Doppler frequency exceeds the Nyquist limit. On the spectral display by Doppler can show peaks going off the display top or bottom, wrapping around the other side of the baseline. On the color display an immediate change in color from one Nyquist limit to the other is seen. **Comet tail** is a reverberation artifact produced when two or more strong reflectors are close together and have a high propagation speed. Then, sound does not travel directly to a reflector and back to the transducer; a strong linear echo appears at the reflector and extends deeper than the reflector. **Enhancement** is an increased relative amplitude of echoes caused by an intervening structure of low attenuation. **Focal enhancement or focal banding** is increased intensity in the focal region that appears as a brightening of echoes shown on the display. **Mirror imaging artifact** is commonly seen around the diaphragm; this artifact results from sound reflecting off another reflector and back. **Mirroring** is the appearance of artifacts on a spectral display when there is improper separation of forward and reverse signal processing channels. Consequently, strong signals will mirror into the other. **Multi-path positioning** and **refraction** artifacts takes place when paths to and from a reflector are different. The longer the sound takes traveling to or from a reflector, the greater the axial error (increased range) in reflector positioning. Refraction and multi-path positioning errors are normally relatively small, contributing to general image degradation rather than to gross errors in object location. **Propagation speed errors** occur when the assumed value for propagation speed by the ultrasound system is incorrect. If the actual speed is greater than assumed, the calculated distance to a reflector is too small, and the reflector will be displayed too far from the transducer. Speed error can cause a structure to be displayed with incorrect size and shape. **Range ambiguity** occurs when reflections are received after the next pulse is transmitted. In ultrasound imaging, it is assumed that for each pulse produced, all reflections are received before the next pulse is sent out. The ultrasound system calculates distance to a reflector from the echo arrival time assuming all echoes were generated by the last emitted pulse. The maximum depth to be imaged unambiguously by the system determines its maximum pulse repetition frequency. **Reverberation** is the continuing reception of a particular signal because of reverberation rather than reflection from a particular acoustic interface. Reverberations are easily identifiable, because they are equally spaced on the display. **Scattering** is the diffuse, low-amplitude sound waves that occur when acoustic energy reflects off tissue interfaces smaller than a wavelength. In diagnostic ultrasound, Doppler signals come primarily from acoustic energy back-scattered from red blood cells. **Shadowing** is the reduction in echo amplitude from reflectors that lie behind a strongly reflecting or attenuating structure. This phenomenon occurs when scanning a lesion or structure with an attenuation rate higher than that of the surrounding tissue. The lesion causes a decrease in beam intensity, which results in decreased echo signals from the structures beyond the lesion. Consequently, a dark cloud behind the lesion image forms on the display. This cloud, or shadow, is useful as a diagnostic clue. **Side lobes** (from single-element transducers) and **grating lobes** (from array transducers) cause objects that are not directly in front of the transducer to be displayed incorrectly in lateral position. **Speckle** artifacts appear as tissue texture close to the transducer but does not correspond to scatterers in tissue. It is produced by ultrasound wave interference and results in general image degradation. **Spectral broadening** is a display phenomenon that occurs when the number of energy-bearing Fourier frequency components increases at any given point in time. As a consequence, the spectral display is broadened. Spectral broadening can indicate disturbed flow caused by a lesion, and it is important diagnostically. However, broadening can also result from interaction between flow and sample volume size. **Speed of sound** artifacts occur if the sound propagation path to a reflector is partially through bone, where sound speed is greater than in average soft tissue. Reflectors appear closer to the transducer than their actual distance because of this greater speed of sound, resulting in a shorter echo transit time than for paths not containing bone.

Acoustic Artifacts in 3D Imaging

Acquisition, rendering, and **editing artifacts** are specific to 3D volume images. Acquisition artifacts are related to patient motion, organ motion, or position-sensing errors. Rendering artifacts include elimination of structures by limiting the region of interest boundaries, thresholding that eliminates structures, and adjacent structure artifacts that add additional information or hide structures. Editing artifacts result from data deleted from a rendered image. **Color** and **Color Power Angio** artifacts include a color flash artifact occurs when gain is set high and the transducer or patient moves. When gain is set too high, the color ROI box fills with color flash. When gain is set low, color bleed can occur. When gain is set too low, insufficient color data renders the image undiagnosable. **Color gain, directional,** and **motion artifacts** occur in 3D imaging. Color gain artifacts are mainly related to the use of excessive gain resulting in random color patterns in 3D images that might be interpreted as diagnostically significant. Directional artifacts are due to aliasing or directional confusion. The velocity range must be set properly, and the relationship between the transducer orientation and the flow vector must be understood. Patient motion can produce flash artifacts that are less obvious in 3D images than in 2D. **Dropout** and **shadowing** are present in 3D imaging although they are more difficult to recognize due to different and unfamiliar displays. Acoustic shadowing and other artifacts look very different when displayed in 3D volumes and may be more difficult to recognize than on standard 2D imaging. Those artifacts may produce apparent defects, such as nonexistent limb abnormalities or facial clefts. Acquiring data from multiple orientations may these artifacts. Fetal limb deficit artifacts are specific to 3D volume images. One explanation for the missing limbs is shadowing caused by adjacent skeletal structures. Overcoming the limb deficit artifact can be accomplished by changing the transducer position and the acquisition plane, as usual. **Motion artifacts** in 3D volumes can be caused by patient motion, fetal movement, cardiac motion, and other movement. Patient motion can produce flash artifacts that are more obvious in 3D images than in 2D. **Pseudoclefting** and **artifacts** are similar to limb deficit artifacts. Artifacts may be present in 3D imaging of the fetal face. As with 2D imaging, it is important to verify putative physical defects by using additional images and other modalities. **Resolution, attenuation,** and **propagation artifacts** are common to 3D imaging. Careful scrutiny of original 2D images is necessary to identify/preclude these artifacts from 3D volume image."

☆☆☆ THE EFFECTIVENESS OF APPROXIMATE MEASUREMENTS AND MODELS

Frederick Mosteller's brilliant essay on approximate vs. refined measurements – and their relation to policy interventions.

"It is the experience of statisticians that when fairly 'crude' measurements are refined, the change more often than not turns out to be small. Merely counting laboratories in a school system is a crude measurement. It is possible to learn more about the quality of laboratories [and test our skepticism about the original crude measurements].

But statisticians would not leap too readily to that... Sadly, in real life the similarities of basic categories are often far more powerful and important than the nice differences which can come to absorb individuals so disposed, but which really don't make a great difference in the aggregate. Statisticians would wholeheartedly say make better measurements, but they would often give a low probability to the prospect that finer measures would produce data leading to different policy. The reasons are several. One is that policy decisions are rather insensitive to the measures – the same policy is often good across a great variety of measures. Secondly, finer measures are something like weights. For example, perhaps one science laboratory is only half as good as another – well and good, let us count it 1/2. It turns out as an empirical fact that in a variety of occasions, we get much the same policy decisions in spite of weights. So there are technical reasons for thinking that finer measurement may not change the main thrust of one's policy. None of this is an argument against getting better information if it is needed, or against having reservations. More data cost money, and one has to decide where good places are to put the next money acquired for investigations. If we think it matters a lot by all means let us measure it better.

Note distinctions between statistical lives and individual lives. For schools, we may not know what works well for specific students, but we do know that if we stopped teaching algebra, few people would ever learn algebra.

Another point about aggregative statistics is worth emphasizing for large social studies. Although the data may not be adequate for decisions about individual persons, they may well be adequate for deciding policy for groups. We may not be able to predict which ways of teaching spelling will be preferable for a given child, but we may be able to say that, on the average, a particular method does better. And then the policy is clear, at least until someone learns how to tell which children would do better under the differing methods."

APPROXIMATE MODELS

Far better an approximate answer to the right question, which is often vague,
than an exact answer to the wrong question, which can always be made precise. JOHN TUKEY

Simple methods typically yield performance almost as good as more sophisticated methods,
to the extent that the difference in performance may be swamped by other sources of uncertainty
that generally are not considered in the classical supervised classification paradigm. DAVID HAND

☆ PRIOR TO DATA ANALYSIS, CONDUCT AN INDEPENDENT FORENSIC DATA AUDIT OF YOUR DATA. DON'T JUST LOOK AROUND, INSTEAD SEE EVERYTHING

If the truth matters, spreadsheets require unbiased forensic audits. Audits are sometimes deflected because researchers and sponsors are anxious to get a peek at the findings early on, despite the well-known anchoring bias that early information has undue influence. Chris Groskopf's *Guide to Bad Data* reveals 46 data quality issues in spreadsheets, a good start for forensic audits.

"Values are missing   Zeros replace missing values   Data are missing you know should be there
Rows or values duplicated   Spelling inconsistent   Name order inconsistent   Date formats inconsistent
Units are not specified   Categories badly chosen   Field names ambiguous   Provenance not documented
Suspicious values present   Data too coarse   Totals differ from published aggregates
Spreadsheet has 65,536 rows   Spreadsheet has 255 or 256 columns   Spreadsheet has dates 1900, 1904, 1969, 1970
Text has been converted to numbers   Numbers have been stored as text
Text garbled   Line endings garbled   Data in a PDF   Data too granular   Data entered by humans
Data intermingled with formatting and annotations   Aggregations computed on missing values
Sample not random   Margin-of-error too large   Margin-of-error unknown   Sample biased
Data have been manually edited   Inflation skews data   Natural/seasonal variation skews data
Timeframe manipulated   Frame of reference manipulated   Author untrustworthy
Collection process opaque   Data assert unrealistic precision   Inexplicable outliers
Data aggregated to wrong categories or geographies   Results p-hacked   Benford's Law fails
An index masks underlying variation   Data are in scanned documents   Too good to be true"

⚠ "DATA CLEANING" IS NOT A FORENSIC AUDIT

Data cleaning programs correct logical inconsistencies, data duplications, impossible values, conflicting postal codes, outliers, other low-tide stuff. A well-designed cleaning program might identify 70% of Groskopf issues, but will have difficulties detecting systemic biases, falsification, and too-good-to-be true – which require experience and honest judgement. A virtue of AI forensic audits is their independent outsider status, unlike the ultimate insider Dr. Confirmation Bias.

☆ GET IT RIGHT FROM THE START IN BUILDING DATABASES: EARLY ADJUSTMENTS ARE CHEAP, LATER RESCUE REVISIONS COSTLY. EXPERT PRACTICAL ADVICE BY JEFF ATWOOD AND NATE SILVER

"Of all the technical debt you can incur, the worst in my experience is bad names – for database columns, variables, functions, etc. Fix those IMMEDIATELY before they metastasize all over your code base and become extremely painful to fix later. . . and they always do." JEFF ATWOOD

"Going through some old data/code. One thing I've learned is when combining different datasets or doing complicated data processing, it pays to be compulsive about missing data or data that doesn't pass sanity checks. More often than you might think, the missing/miscoded/outlier cases indicate a larger, more systematic problem with your code or data. My advice is to do due diligence before moving onto next steps, because errors tend to compound." NATE SILVER

"Bioinformatics . . . or 'advanced file copying'" NICK LOMAN

SPREADSHEET AUDITS ARE ESSENTIAL

⚠ PROVENANCE IS NOT DOCUMENTED

"Data are created by businesses, governments, nonprofits, nut-job conspiracy theorists. Data are gathered in different ways – surveys, sensors, satellites. Knowing where data came from provides huge insights into its limitations. Survey data is not exhaustive. Sensors vary in accuracy. Governments are disinclined to give you unbiased information. War-zone data are geographically biased due to dangers of crossing battle lines. Various sources are daisy-chained together. Policy analysts redistribute government data. Every stage in that chain is an opportunity for error. Know where your data came from." Chris Groskopf

⚠ BEWARE OF HUMAN-ENTERED DATA: THE CHIHUAHUA SYNDROME

"There is no worse way to screw up data than to let a single human type it in, without validation. I acquired a complete dog licensing database. Instead of requiring people registering their dog to choose a breed from a list, the system gave dog owners a text field to type into, so this database had 250 spellings of **Chihuahua.** Even the best tools can't save messy data. Beware of human-entered data." Chris Groskopf ⚠ ELECTRONIC HEALTH RECORDS

⚠ REDEFINING GROUPS MAY BREAK RANDOM ASSIGNMENT

Medical researchers sometimes redefine treatment groups to include only *patients treated* instead of *intended to treat*. But consider research randomly assigning patients to treatment vs. placebo: if the treatment takes place on the 5th floor in a building without elevators, for example, then less healthy patients may never arrive for treatment (example by Darrel Francis). Even if the treatment *has no effect at all*, the treated group will now appear to do better than placebo groups.

⊘ HOW DR. CONFIRMATION BIAS MESSES AROUND WITH MEASUREMENTS AND SPREADSHEETS

Effects in current medical research are small, large effects have already been discovered. And thus small pseudo-findings can be fabricated by a few cheats. Dr. Confirmation Bias doing data fudges:

"Try 5 methods to manage missing data, see what works best." "Patient compliance always averages out" "High/Low bins: try cuts at median, mean, midmean, isosceles harmonic mean" "Results not good, transform variables: log, arc sine, trichotomize, deca-chotomize, whatever it takes" "Report summary models only" "Look at absolute rates privately, report relative rates in press releases" "Too late to audit spreadsheet, findings optimized on unaudited data" "Consult with biostat folks at end of our data work, their job is to make our results truly significant: .001 or .01, **not** .05. After all, **our** research grant is paying the biostat guys. Whose side are they on?" "Find all subgroups where our desired findings work best, it's only fair" "Stonewall all requests to see our original data" "If results contrary to our sponsor's expectations: (1) Do not publish, (2) Stop it with randomized trials, (3) Actually, sponsor prefers no control arm at all, with outcomes measured by patient post-op/post-chemo Gratitude and Hope, never measure/report reductions in all-cause mortality."

⚠ BIOSTATISTICIANS DESCRIBE CHEAT REQUESTS MADE BY RESEARCHERS

522 consulting biostatisticians surveyed, and 75% responded. The survey reported these common inappropriate requests by researchers: *"removing or altering some data records to better support the research hypothesis; interpreting the statistical findings on the basis of expectation, not actual results; not reporting the presence of key missing data that might bias the results; ignoring violations of assumptions that would reverse the results."*

"Researcher Requests or Inappropriate Analysis and Reporting: U.S. Survey of Consulting Biostatisticians," Min Qi Wang, Alice F. Yan, Ralph V. Katz, *Annals of Internal Medicine*, 2018, edited

ASSESSING MEASUREMENT QUALITY

☆☆☆ DIRECTLY OBSERVE DATA COLLECTION AT THE EXACT MOMENT OF MEASUREMENT

☆☆☆ MEASUREMENT ERRORS ARE MEASURED BY OBSERVING DATA COLLECTION, DETECTIVE WORK, DATABASE AUDITS – NOT BY STANDARD STATISTICAL ERROR ESTIMATIONS

⚠ BATCH EFFECTS IN SEQUENCING GENOME DATA

"High-throughput technologies are widely used to assay genetic variants, gene and protein expression, epigenetic modifications. Often overlooked complications are *batch effects,* occurring because measurements are affected by laboratory conditions, reagent lots, and personnel differences. This is a major problem when *batch effects are correlated with an outcome of interest and lead to incorrect conclusions.* Batch effects (and other technical and biological artifacts) are widespread and critical to address."

Below, batch effects for 2nd-generation sequencing data from the 1000 Genomes Project. "Each row is a different HapMap sample processed in the same facility/platform. Samples are ordered by processing dates. Coverage data from each feature are standard across samples. Dark blue represents 3 standard deviations below average. Orange represents 3 standard deviations above average. Many batch effects are observed, and the largest one occurs between days 243–251 (long orange horizontal streaks)."

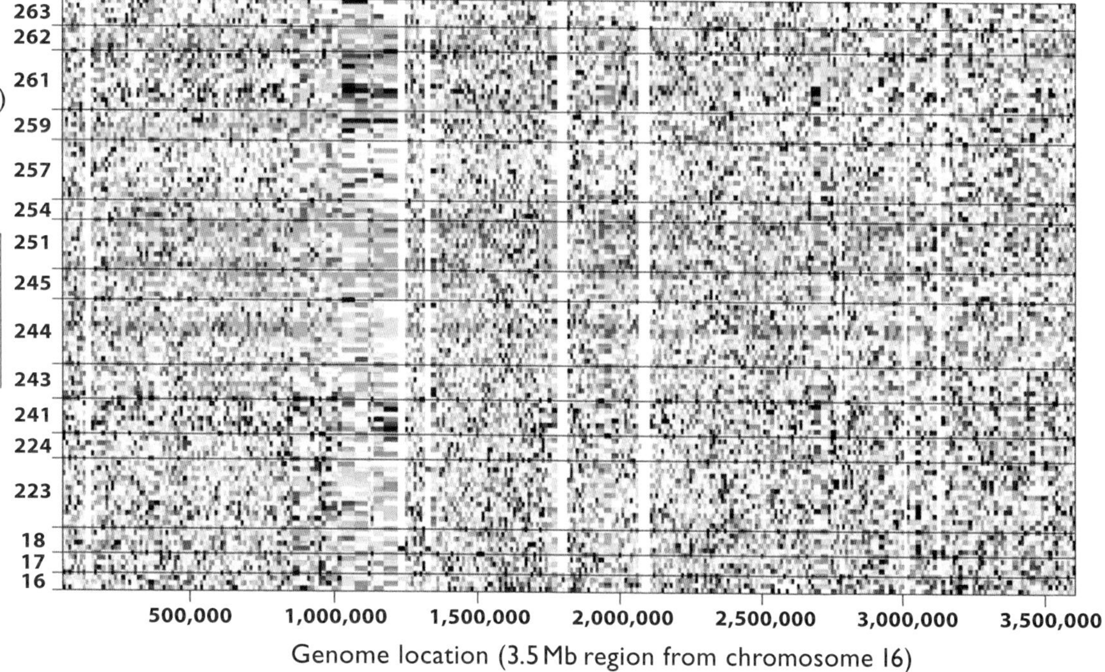

Jeffrey T. Leek, et al, "Tackling the widespread and critical impact of batch effects in high-throughput data," HHS Public Access authors manuscript, 2010, edited; *Nature Reviews Genetics* 11, 2010, 733-739.

⚠ SURROGATE/PROXY/BIOMARKER MEASUREMENTS VS. MEDICAL PATIENT OUTCOMES

Measurements often serve as useful and convenient *surrogates/proxies/biomarkers* – which are based on *evidence/assumption/theory/folklore/marketing/convention* claiming that biomarkers signal actual patient outcomes. For example, markers such as blood pressure, lipids, patient hope and gratitude – rather than serious long-run outcomes such as incidence of strokes, heart attacks, quality of life, all-cause mortality. Only unbiased empirical evidence can assess links between markers and patient outcomes that are relevant, resilient, meaningful. Changing hard outcomes to easier marker goals is a clear signal that the intervention doesn't work very well. For example, a new widely used oncology marker variable is *progression-free survival,* now playing a starring role in FDA Drug Approval Theater:

"Only one-third of cancer drugs entering Europe/U.S. markets have evidence of overall benefits in survival or quality of life. But regulators routinely approve new cancer drugs using surrogate endpoints measuring 'progression-free survival,' as the preferred surrogate endpoint in oncology drug trials, *not patient benefits.* After approval, drug companies have little or no incentive to evaluate the clinical benefit of their products. Data on overall survival or quality of life rarely emerge, even years after market entry. New cancer drugs should be approved on the basis of their overall survival and quality of life benefits."

Christopher M. Booth, Elizabeth A. Eisenhauer, "Progression-Free Survival: Meaningful or Simply Measurable?", *Journal of Clinical Oncology*, 2012; v.12, 10, edited

⚠ "GROUND TRUTH" IS JUST SOMEONE ELSE'S SPREADSHEET

In ordinary language, "ground truth" is information collected on location, but database ground truthers never see their data at the moment of actual measurement. All such databases require independent forensic audits, checking for batch processing errors, changes and biases in measurement practices in space and time, and 100 other issues. Attributing truth to a database and claiming 'proof of concept' fools around with the meanings of truth and proof. This is the Fallacy of Equivocation: 'whenever a term is used in 2 or more senses within a single argument, so that a conclusion appears to follow when it in fact does not,' – as in the puns *ground truth, error, power, optimal, causal model, explained/unexplained variance, intellectual property.* Nature's laws and survival rates are authentic Ground Truths.

⚠ ERRORS IN, ERRORS OUT ⚠ BIAS IN, BIAS OUT ⚠ NO CAUSALITY IN, NO CAUSALITY OUT
⚠ CIRCULARITY IN, CIRCULARITY OUT ⚠ BIG DATA, ML, AI HUBRIS ⚠ FALLACY OF EQUIVOCATION

⚠ SURVIVAL BIAS

Most medieval castles were made of wood. We think most were made of stone because of survivor bias. Research databases are those that survived long enough to be selected for Ground Truth status. Survivor bias, subtle and inscrutable, requires deep meta-cognition and detective work about database provenance.

⚠ IMMORTAL TIME BIAS ⚠ SKETCHY IN, SKETCHY OUT ⚠ BIAS IN, BIAS OUT

⚠ ENSHRINING AND REINFORCING PAST PRACTICES, A POLICE PREDICTION MODEL SUFFERS FROM GARBAGE-IN/GARBAGE-OUT. BUT MODEL OUTPUTS, IN EFFECT, MEASURE THEIR BIASES – AND SO PREDICTION MODELS DO RAT THEMSELVES OUT.

Predictive policing is a widely used model. In daily work, some predictive systems may use "dirty data" that enshrines and intensifies past police practices. These models document the prevalence of unlawful and inefficient practices. There will never be better documentation, for it would be difficult to investigate/gather/model comparable data afresh. *Thus predictive data models directly measure prevailing biases!*

DIRTY DATA, BAD PREDICTIONS: HOW CIVIL RIGHTS VIOLATIONS IMPACT POLICE DATA, PREDICTIVE POLICING SYSTEMS, AND JUSTICE. RASHIDA RICHARDSON, JASON SCHULTZ, KATE CRAWFORD

"Law enforcement agencies are increasingly using algorithmic predictive policing systems to forecast criminal activity and allocate police resources. Yet in numerous jurisdictions, these systems are built on data produced within the context of flawed, racially fraught, and sometimes unlawful practices ('dirty policing'). This can include systemic data manipulation, falsifying police reports, unlawful use of force, planted evidence, unconstitutional searches. These policing practices shape the environment and the methodology by which data is created, which leads to inaccuracies, skews, and forms of systemic bias embedded in the data ('dirty data'). Predictive policing systems informed by such data cannot escape the legacy of unlawful or biased policing practices that they are built on.

Nor do claims by predictive policing vendors that these systems provide greater objectivity, transparency, or accountability hold up. While some systems offer the ability to see algorithms used and even occasionally access to the data itself, there is no evidence to suggest that vendors independently or adequately assess the impact that unlawful and biased policing practices have on their systems, or otherwise assess how broader societal biases may affect their systems.

Confirmation Feedback Loops

Though there is research that empirically demonstrates that the mathematical models of predictive policing systems are susceptible to runaway feedback loops, where police are repeatedly sent back to the same neighborhoods regardless of the actual crime rate, such feedback loops are also a byproduct of the biased police data. More specifically, police data can be biased in two distinct ways. Fundamentally, police data reflects police practices and policies. If a group or geographic area is disproportionately targeted for unjustified police contacts and actions, this group or area will be over-represented in the data, in ways that often suggest greater criminality. Second, the data may omit essential information as a result of police practices and policies that overlook certain types of crimes and certain types of criminals. For instance, police departments, and predictive policing systems, have traditionally focused on violent, street, property, and quality of life crimes. Meanwhile, white collar crimes are comparatively under-investigated and over-looked in crime reports. Available studies estimate that 49% of businesses and 25% of households have been victims of white collar crimes, compared to a 1.1% prevalence rate for violent crimes and 7.4% prevalence for property crime."

New York University Law Review Online February 13, 2019, edited.

DATA MODELS CONTAIN BOTH REAL AND IMAGINARY PARTS

Millions of successful models thrive in physics, chemistry, engineering– because mathematical laws define ground truth. Human behavior research lacks such assurances: instead of Nature's laws, we are stuck with after-the-fact statistical modeling. These models can produce handwaving, patent claims, disciplinary cults. How do we know if models are true and work? Models require experimental tests and applied interventions to learn the truth, just like real science:

'Sailors talk about hydrodynamics the way CEOs talk about macroeconomics: they either treat it with mystical reverence, or they claim to understand it and are wrong. Unlike macroeconomics, though, if you know what you are doing you can test the propositions of hydrodynamics on actual physical models in a lab.' BRENDAN GREELEY, FINANCIAL TIMES, MARCH 25, 2021

STATISTICAL MODELS: USES, ASSUMPTIONS, ADVERSE REACTIONS

Detailed package-inserts accompany prescription drugs. Consider a similar document for statistical models – describing use, pitfalls, prevalence of adverse effects, model breakdowns. *Imagine if encounters with every statistical model – in textbooks, computer code, workaday practice, publications – had to face up to adverse reactions. After all, these models can affect thousands of statistical lives.* At left, the first of 16 pages devoted to LISINOPRIL, a cardiovascular drug. At right, this mock-up document alerts users to model assumptions, breakdowns, adverse reactions – for a widely-used statistical model, stepwise logistic multiple regression

ADVERSE EFFECTS LISINOPRIL

The incidence of adverse effects varies according to the disease state of the patient:

People taking lisinopril for *treatment of hypertension* may experience the following side effects:

Headache (3.8%) Dizziness (3.5%) Cough (2.5%)
Difficulty swallowing or breathing (signs of angioedema) allergic reaction (anaphylaxis)
Hyperkalemia (2.2% in adult clinical trials)
Fatigue (1% or more) Diarrhea (1% or more)
Some severe skin reactions have been reported rarely, including toxic epidermal necrolysis and Stevens-Johnson syndrome; causal relationship has not been established.

People taking lisinopril for treatment of *myocardial infarction* may experience the following side effects:

Hypotension (5.3%) Renal dysfunction (1.3%)

People taking lisinopril for the treatment of *heart failure* may experience the following side effects:

Dizziness (12% at low dose – 19% at high dose)
Hypotension (3.8%) Chest pain (2.1%)
Fainting (5-7%)
Hyperkalemia (3.5% at low dose – 6.4% at high dose)
Difficulty swallowing or breathing (signs of angioedema), allergic reaction (anaphylaxis)
Fatigue (1% or more) Diarrhea (1% or more)
Some severe skin reactions (toxic epidermal necrolysis, Stevens-Johnson syndrome) have been reported rarely, causal relationship not established.

BLACK BOX WARNING !
STEPWISE LOGISTIC MULTIPLE REGRESSION:
DO NOT USE FOR MAKING CAUSAL INFERENCES.
APPROVED ICU USE: COMPASSIONATE THERAPY
FOR STAGE 3 DUSTBOWL EMPIRICISM

ADVERSE EFFECTS
STEPWISE LOGISTIC MULTIPLE REGRESSION (SLMR)
(TRADE NAMES: MACHINE-LEARNING, ML, AI)

SLMR-ML data analysts may experience adverse outcomes:
Cherry-picked models (80.1%) Model and p-hacking (>70%)
Ioannidis syndrome: most published medical research is false
Unpublishable by highly-ranked medical journals (40.0%)
Dizziness (23%) Fatigue (11.8% or more) Fainting (7.3%)
Table 2 fallacy: confounding direct and indirect causes (100%)
Multicollinearity (80.5%) Hankins Condition: 'A journey of a thousand hypotheses begins with a single SLMR' Failure to replicate (70.9%) Accusations of garbage in, garbage out.

Model assumptions: one-way independent causes, errors independently and identically distributed, x-variables measured without error. Feedback, simultaneity, interaction effects assumed not to exist; or, if modeled, more assumptions and more data are needed. Implicitly assumes SLMR finds the single best equation, practical experience suggests there are better models with different variables.

Daniel Westreich and Sander Greenland, "The Table 2 Fallacy," *Am J Epidemiol*, 2013, 177, 4, 292–298; Cuthbert Daniel and Fred Wood, *Fitting Equations to Data*, 1980, 84–85; Gary Smith, "Step away from stepwise," *Big Data*, 2018, 5:32.

☆☆☆ EXEMPLARY REPORT OF POSSIBLE SIDE-EFFECTS OF KAPLAN-MEIER SURVIVAL ESTIMATES

Kaplan-Meier curves track survival times, numbers of patients living over a period of time after a medical intervention. The abstract of this famous paper (>50,000 citations) warns that "lifetime (age at death) is independent of potential loss time; in practice this assumption deserves careful scrutiny." KM lines show data directly, and the quality of inference depends on the character of the data. Engineers at JnF Practical Quality Control have produced an excellent package insert on issues in KM survival times. *Each use of a model should remind users of constraints, assumptions, breakdowns.*

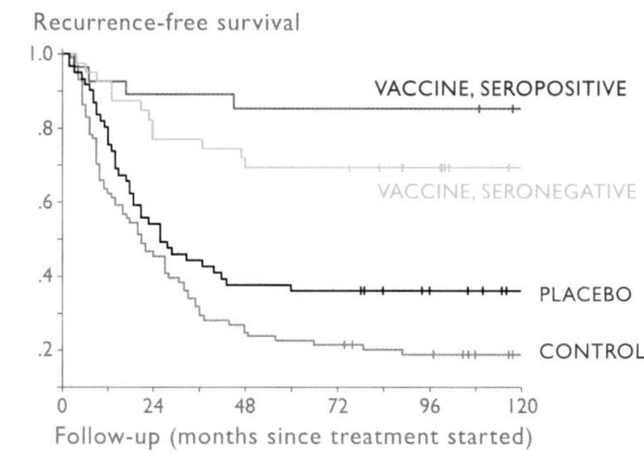

"**Implicit factors** Lack of independence within a sample is often caused by an implicit factor in the data. For example, if we are measuring survival times for cancer patients, diet may be correlated with survival times. If we do not collect data on implicit factors (diet in this case), and the implicit factor has an effect on survival times, then we in effect no longer have a sample from a single population, but instead a sample that is a mixture drawn from several populations, one for each level of the implicit factor, each with a different survival distribution. Implicit factors affect censoring times, by affecting the probability that a subject will be withdrawn from the study or lost to follow-up. For example, younger subjects may move away (and be lost to follow-up) more frequently than older subjects, so that age (an implicit factor) is correlated with censoring. If the sample under study contains many younger people, the results of a study may be substantially biased because of different patterns of censoring. This violates the assumption that censored values and noncensored values all come from the same survival distribution. Stratification can control for an implicit factor.

Lack of independence of censoring If pattern of censoring is not independent of survival times, then survival estimates may be too high (if subjects who are more ill tend to be withdrawn from the study), or too low (if subjects who will survive longer tend to drop out of the study, lost to follow-up). The estimates for the survival functions and their variances rely on independence between censoring times and survival times. If independence does not hold, the estimates may be biased, and the variance estimates may be inaccurate. An implicit factor not accounted for by stratification may lead to a lack of independence between censoring times and observed survival times.

Lack of uniformity within a time interval Kaplan-Meier estimates for survival functions and standard errors rely on assumptions that the probability of survival is constant within each interval (although it may change from interval to interval), where the interval is the time between two successive noncensored survival times. If the survival rate changes during the course of an interval, then the survival estimates for that interval will not be reliable or informative.

Many censored values A study may end up with many censored values, from having large numbers of subjects withdrawn or lost to follow-up, or from having the study end while many subjects are still alive. Large numbers of censored values decrease the equivalent number of subjects exposed (at risk) at later times, making Kaplan-Meier estimates less reliable than they would be for the same number of subjects with less censoring. Moreover, if there is heavy censoring, the survival estimates may be biased (because the assumption that all censored survival times occur immediately after their censoring times may not be appropriate), and estimated variances become poorer approximations, perhaps considerably smaller than the actual variances. A high censoring rate may also indicate problems with the study: ending too soon (many subjects still alive at the end of the study), or a pattern in censoring (many subjects withdrawn at the same time, younger patients lost to follow-up sooner than older ones). If the last observation is censored, the Kaplan-Meier estimate of survival can not reach 0.

Patterns in plots of data If the assumptions for the censoring and survival distributions are correct, then a plot of either the censored or the noncensored values (or both together) against time should show no particular patterns, and the patterns should be similar across the various groups.

group 1
group 2
time

Special problems with small sample sizes Time intervals in Kaplan-Meier calculations are determined by distinct noncensored survival times. This means that the smaller the sample size, the longer the intervals will be, raising questions of whether the assumption of a constant survival probability within each interval is appropriate. Small samples make it difficult to detect possible dependencies between censoring and survival, or the presence of implicit factors. If the number of subjects exposed (at risk) in an interval or the number of subjects that survived to the beginning of that interval is small, variance estimates for survival functions will tend to underestimate actual variance. This situation is most likely to occur for later intervals, when most subjects have either died or been censored, so that variance estimates for later intervals are less reliable than those for earlier intervals."

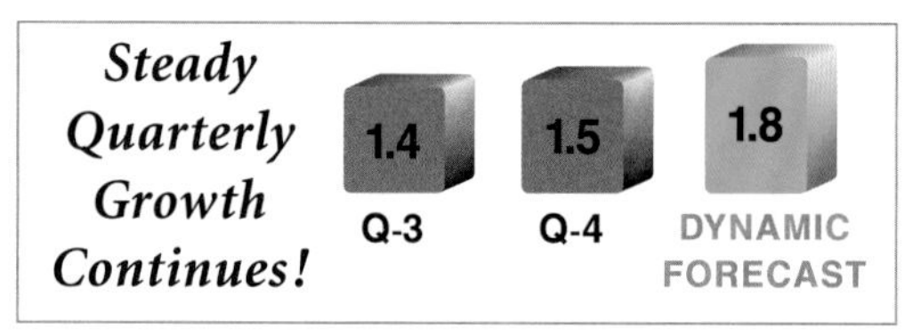

Summary reports are economical with the truth: this slide cherry-picks, over-summarizes the data – and disrespects the audience. *"The Board of Directors loves good news. Our internal numbers are not entirely good news. Need to fix it in our slides. Everyone does it."*

Recent Mediocrity to Continue!

Q-1	Q-2	Q-3	Q-4	FORECAST
4.0	4.6	1.4	1.5	1.8

Two additional quarters of data change the story. How about 12 more quarters? But regardless of how many quarters, readers are shown *binned quarterly data,* just one number per quarter. LittleDataGraphics are friends of falsity, enemies of truth.

Recent mediocrity accompanied by high volatility. In Q1-4, possible accounting manipulations. Oops!

5 4 3 2 1

Q-1 Q-2 Q-3 Q-4

Detailed data moves closer to the truth. No more binning, less cherry-picking, less truncation. This graphic reveals increases in recent volatility. End-of-quarter-upticks may signal accounting manipulation (premature revenue recognition). This time-series is easily readable by all – Boards of Directors, financial journalists, shareholders, colleagues. The initial slide 1.4 1.5 1.8 treats viewers like mushrooms kept in the dark in manure.

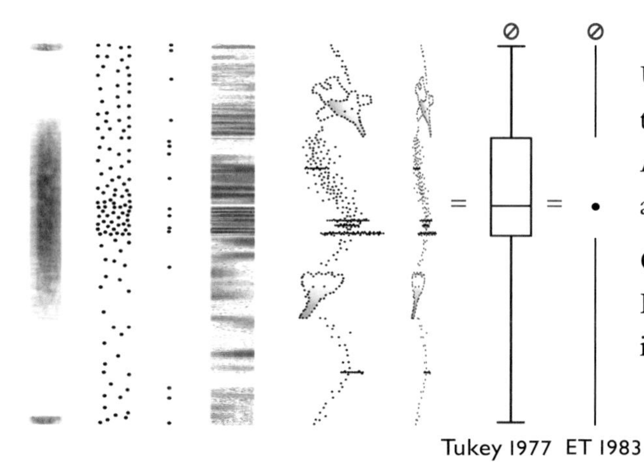

Upon learning about averages, every schoolchild knows that divergent data sets can produce identical summaries. At left, 6 different data sets and their 63 possible combinations all yield the same identical boxplot.

Censoring data often produces false findings. Show the data. Nowadays display screens have enormous resolution. The 2021 iPhone screen shows 3,566,952 pixels, 458 per square inch.

Can boxplots chase down outliers, as *Computational Methods in Physics* (2018) claims? Why show only the 2 extreme min/max outliers and bin all the other data? Unlike this LittleData boxplot, an unbinned plot of the same data shows all 1000 measurements:

Now the 2-dimensional case:

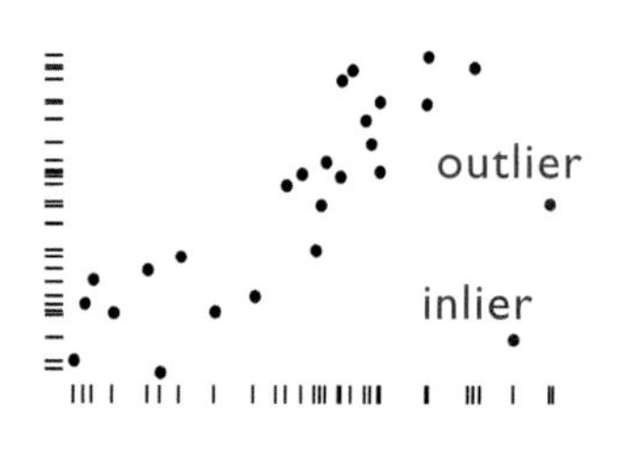

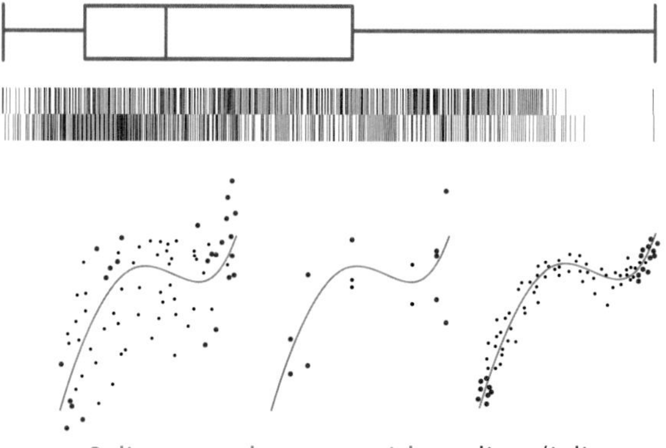

like boxplots, glib published summary models lack credibility

3 divergent data sets with outliers/inliers, but the same naive summary model describes all 3 different data sets!

BINNING CONTINUOUS DATA IS MEDIUM-QUALITY EVIDENCE OF FALSIFICATION

Multiple boxplots cheat. Golden boxplots double-bin the flying elephants data set: *X* into 6 columns, *Y* into quartiles/outliers – making false inferences. Binning creates thresholds/plateaus notoriously difficult to replicate or verify in research on humans. Thresholds are model-imposed, not data-driven. Some medical research reports are written by the sponsor's ghostwriters who spin data into a product pitch. Sponsors can suppress the publishing of research contrary to desired results. Publication bias corrupts the collective knowledge (meta-analysis) of entire research fields.

ORIGINAL XY-DATA
12 elephants flying in sinusoidal flight-paths with identical local pulses. N = 2,598

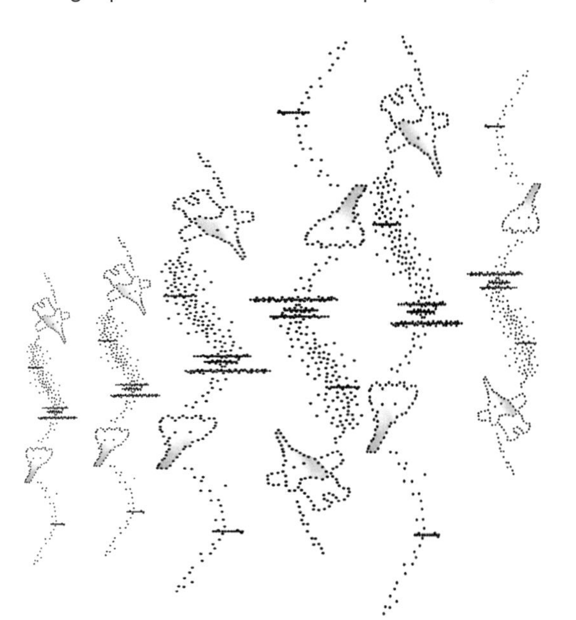

BOXPLOTS = FALSE FINDINGS
"As X dosage increases, response Y makes steady gains to reach a plateau – a highly significant and novel threshold – that signals need for increased dosages of drug X." N = 2,598

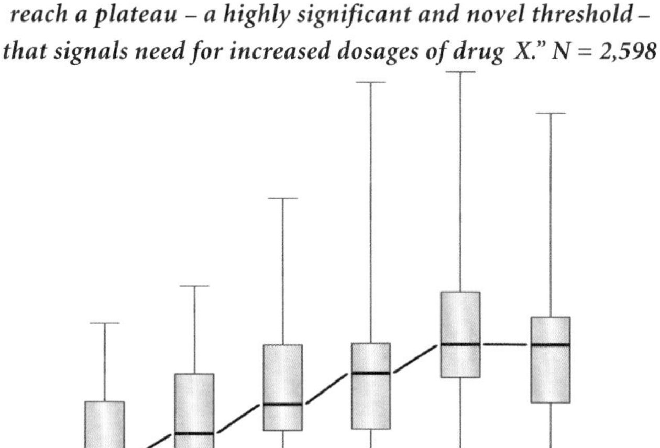

⊘ BOXPLOT CHEATS ⊘ DOUBLE-BINNING ⊘ IMAGINARY THRESHOLDS/PLATEAUS/CUT-POINTS ⊘ SPINNING
⊘ GHOSTWRITING, GHOST DATAVIZ ⊘ PUBLICATION BIAS ☆ FOLLOW THE MONEY ☆ SHOW THE DATA

DATA GRAPHICS AND DATA AVAILABILITY: HOW DR. CONFIRMATION BIAS JUSTIFIES CHEATING

"Smoothed data summaries reduce clutter, make our results understandable to journalists/doctors/sponsors. Readers don't want *data data data!* They love simple graphics with a strong message. This isn't rocket science. Frankly, readers look only at abstracts, graphics, citations – so that's where our team works hard to pitch our findings. Reasons for not making our data available: Trade secret. Violate patient privacy. Hard drive crashed. Intellectual property. In litigation. Double top secret. Patent pending. IPO silent period. All of the above."

REPLY TO DR. CONFIRMATION BIAS: STOP CHEATING

"Clutter" in data graphics is evidence that your models don't fit the data – and that you know it. You also know that your summary graphics cover up contrary data and depict dubious thresholds not present in the data set. Such cheats are obvious and easily detected, and damage your credibility.

On high-resolution data: every day a billion people look at e-maps with data densities 20 times greater than your deceptive LittleDataGraphics. Mapmakers and scientists publish readable graphics showing immense data. Why assume that readers suddenly become stupid just because they're reading your research report? To improve learning from data, credibility, and integrity, show the data.

Evidence for a large exomoon orbiting Kepler-1625b

Alex Teachey and David M. Kipping ***Science Advances 4,10, 03 October 2018, edited.***

"Exomoons are the natural satellites of planets orbiting stars outside our solar system. We present new observations of a candidate exomoon associated with Kepler-1625b using the Hubble Space Telescope to validate or refute the moon's presence. We find evidence in favor of the moon hypothesis, based on timing deviations and a flux decrement from the star consistent with a large transiting exomoon. Self-consistent photodynamical modeling suggests that the planet is likely several Jupiter masses, while the exomoon has a mass and radius similar to Neptune. Since our inference is dominated by a single but highly precise Hubble epoch, we advocate for future monitoring of the system to check model predictions and confirm repetition of the moon-like signal."

This paper exemplifies the authentic presentation of data and uncertainties. The summary is written in straight-forward language describing the findings and need for replication. Unlike many medical research publications, this paper is not paywalled; computer code and curve-fitting are open-source and replicable, not absurdly claimed to be intellectual property, and the exomoon is not patented, trademarked, lawyered up, monetized.

These exoplanet data graphics show thousands of measurements along with 3 competing models (linear, quadratic, exponential) from 2 data sources – with the resolution and readability worthy of Google maps. These data displays show sensitivity tests of the statistical models and their data sources:

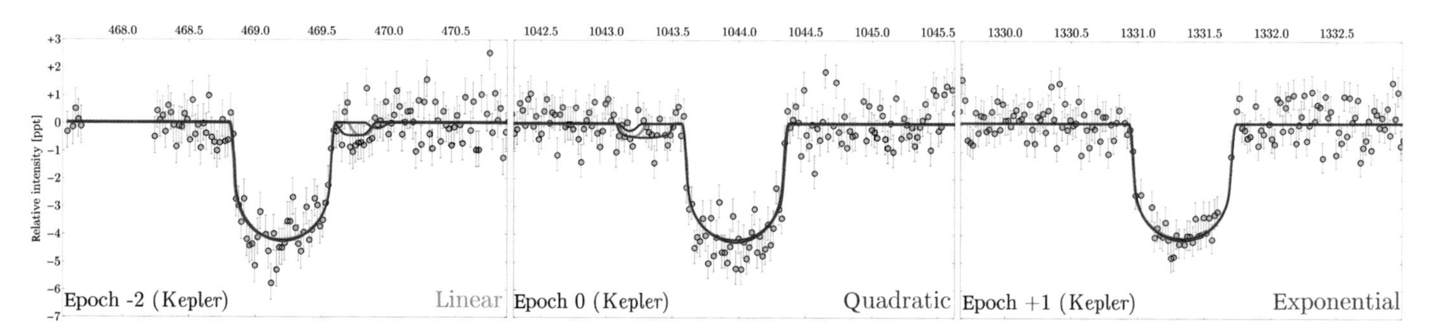

Moon solutions. The three transits in Kepler (top) and the October 2017 transit observed with HST (bottom) for the three trend model solutions. The three colored lines show the corresponding trend model solutions for model M, our favored transit model. The shape of the HST transit differs from that of the Kepler transits owing to limb darkening differences between the bandpasses.

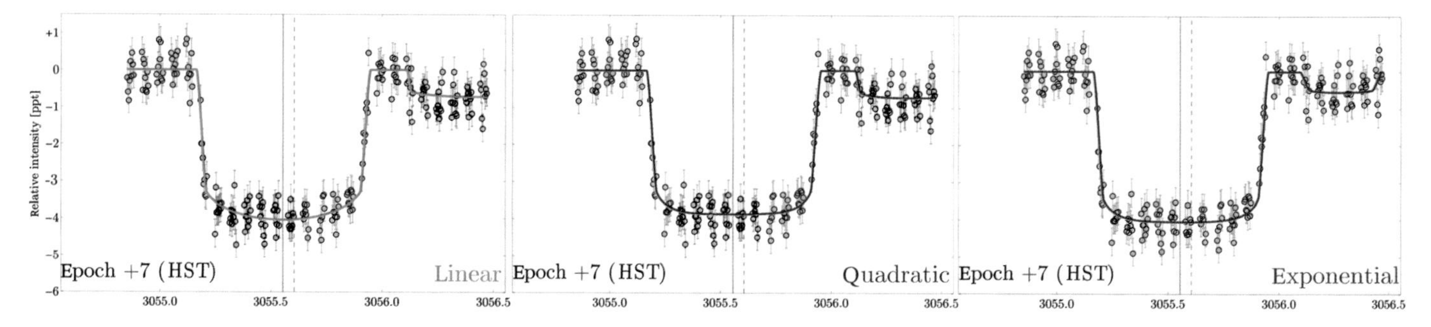

Alex Teachey and David M. Kipping, "Evidence for a large exomoon orbiting Kepler-1625b"
Science Advances 4, 10, 03 October 2018

MAKING JUDGEMENTS ABOUT UNCERTAINTY, CREDIBILITY, TRUTH IN EMPIRICAL RESEARCH

Uncertainties are inherent in data analysis. Uncertainties and errors are entangled with one another and just about anything else that moves – contrary to standard statistical models, where a few over-modeled numbers based on empirically false assumptions create an illusion of certainty. Rather, the credibility of evidence-based conclusions requires detective work about measurements, how those measurements were analyzed, the character and quality of the substantive explanatory theory and its competitors, and avoiding pitfalls. The exomoon research report concludes with exemplary judgements of uncertainty. It is certain that exoplanets exist (~5000 already identified), and moons accompany planets (~200 moons in our solar system). But is the model describing a *real* exomoon or creating an *illusory* exomoon – perhaps due to anomalies in signal processing, or perhaps it is an exo*planet,* not an exo*moon?*

Together, a detailed investigation of a suite of models tested in this work suggests that the exomoon hypothesis is the best explanation for the available observations. The two main pieces of information driving this result are (i) a strong case for TTVs [transit timing variations] in particular a 77.8-min early transit observed during our HST observations, and (ii) a moon-like transit signature occurring after the planetary transit. We also note that we find a modestly improved evidence when including additional dynamical effects induced by moons aside from TTVs.

3 counter-explanations are countered by specific evidence. Are there other counter-explanations, known or unknown?

The exomoon hypothesis is further strengthened by our analysis that demonstrates that (i) the moon-like transit is not due to an instrumental common mode, residual pixel sensitivity variations, or chromatic systematics; (ii) the moon-like transit occurs at the correct phase position to also explain the observed TTV; and (iii) simultaneous detrending and photodynamical modeling retrieves a solution that is not only favored by the data but is also physically self-consistent.

A thoughtful sentence, then followed up by acknowledging three remaining issues.

Together, these lines of evidence all support the hypothesis of an exomoon orbiting Kepler-1625b. The exomoon is also the simplest hypothesis to explain both TTV and post-transit flux decrease, since other solutions would require two separate and unconnected explanations for these two observations. There remain some aspects of our present interpretation of the data that give us pause. First, the moon's Neptunian size and inclined orbit are peculiar, though it is difficult to assess how likely this is *a priori* since no previously known exomoons exist. Second, the moon's transit occurs toward the end of the observations and more out-of-transit data could have more cleanly resolved this signal. Third, the moon's inferred properties are sensitive to the model used in correcting Hubble Space Telescope's visit-long trend, and thus some uncertainty remains regarding the true system properties. However, the solution we deem most likely, a linear visit-long trend, also represents the most widely agreed upon solution for the visit-long trend in the literature.

Formal models yield a goes-to-eleven Bayes factor, but then followed up by list of concerns that can't be quantified.

In reasoning about uncertainty, *known* and *unknown unknowns* are surely the case, and are acknowledged here, and should be similarly acknowledged in medical research publications.

All in all, it is difficult to assign a precise probability to the reality of Kepler-1625b-i [the possible exomoon]. Formally, the preference for the moon model over the planet-only model is very high, with a Bayes factor exceeding 400,000. But, this is a complicated and involved analysis where a minor effect unaccounted for, or an anomalous artifact, could potentially change our interpretation. In short, it is the unknown unknowns that we cannot quantify. These reservations exist because this would be a first-of-its-kind detection—the first exomoon. Historically, the first exoplanet claims faced great skepticism because there was simply no precedence for them. If many more exomoons are detected in the coming years with similar properties to Kepler-1625b-i exomoon, it would hardly be a controversial claim to add one more."

Alex Teachey and David M. Kipping, "Evidence for a large exomoon orbiting Kepler-1625b," *Science Advances* 4, 10, 03 October 2018, edited.

Flying at high altitude over a crime scene, standard statistical analysis is based on puns – 'error' 'confidence' 'unexplained variance' 'significant' 'power' 'independent' – *but empirical uncertainties and errors are not detected/measured/modeled by standard statistical methods.* It is falsification to estimate 'uncertainties' under false assumptions, announce a hypothesis is 'true' or a numerical difference is 'significant'. Standard model assumptions, necessary for mathematical tractibility, are briefly described in textbooks, then forgotten/hidden in workaday data analysis and published reports.

Are assumption-bound classical error models too certain and exact about too many things – and thus too certain and exact for evaluating empirical uncertainty? Links between math models noisy reality are created by punning, calling two different things by the same name. Models also include information about the attitudes of researchers and even readers, although the data doesn't care. These ambiguities/puns produced 100 years of insider epistemological debates about meaning of statistical credibility among readers, researchers, journal editors, statistical societies. Leave epistemology to the Departments of Philosophy and History of Science.

CREDIBLE DATA-BASED CONCLUSIONS

In Pandemic month 5, investigator-initiated research by the Oxford Randomized Evaluation of Covid-19 Group found dexamethasone reduced mortality among critically-ill patients:

> 'In patients hospitalized with COVID-19, dexamethasone reduced 28-day mortality among those receiving invasive mechanical ventilation or oxygen at randomization, but not among patients not receiving respiratory support. Dexamethasone reduced deaths by one-third in patients receiving invasive mechanical ventilation (29.0% vs 40.7%, RR 0.65), and by one-fifth in patients receiving oxygen without invasive mechanical ventilation (21.5% vs 25.0%, RR 0.80), but did not reduce mortality in patients not receiving respiratory support at randomization (17.0% vs 13.2%, RR 1.22).'

'For less than £50 (US$63), you can treat eight patients and save one life.' Dexamethasone is already available worldwide. (Given an actual cost of $63, U.S. hospitals will likely try to collect $800 to $5,000.)

This study is timely and smart: a randomized controlled trial, many UK sites, *all-cause mortality* is the measured outcome (not proxies, surrogates, markers), a sufficient *n*, relative/absolute risks shown together. This rapid RCT platform then assessed many other covid-19 interventions.

Few financial conflicts for these researchers: they do public health, not profits. Covid-19 treatments are difficult enough, without limiting research solely to new patented drugs.

Often first-discovery evidence is the most enthusiastic that will ever be found, *too good to be true*. Independent replications are a necessity. Ten weeks later, the World Health Organization published meta-analyses of 6 small new RCTs + the original study, now with data from 12 countries – and replicated the original findings, and extended the results to other corticosteroids.

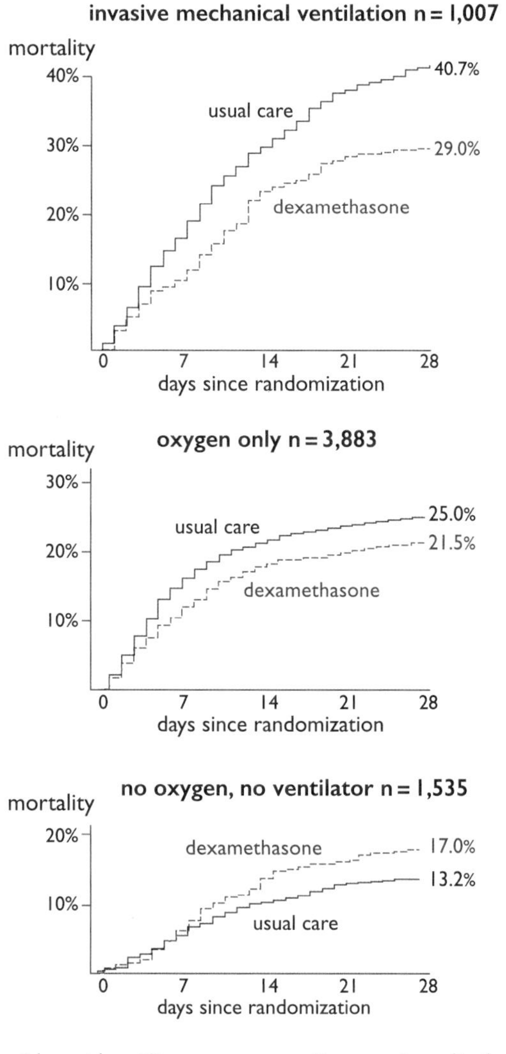

Peter W Horby, Martin J Landray, et al, 'Effect of dexamethasone in hospitalized patients with covid-19,' June 22, 2020, online preprint, edited; Kai Kuperschmidt, 'One U.K. trial is transforming COVID-19 treatment. Why haven't others delivered more results?' *Science,* July 2, 2020. 'Association Between Administration of Systemic Corticosteroids and Mortality Among Critically Ill Patients With COVID-19,' meta-analysis by WHO Rapid Evidence Appraisal for COVID-19 Therapies (REACT) Working Group, *JAMA,* published online September 2, 2020.

MODEL MULTIPLICITY: SAME DATA, BUT DIFFERENT MODELS, MODELERS, MESSAGES

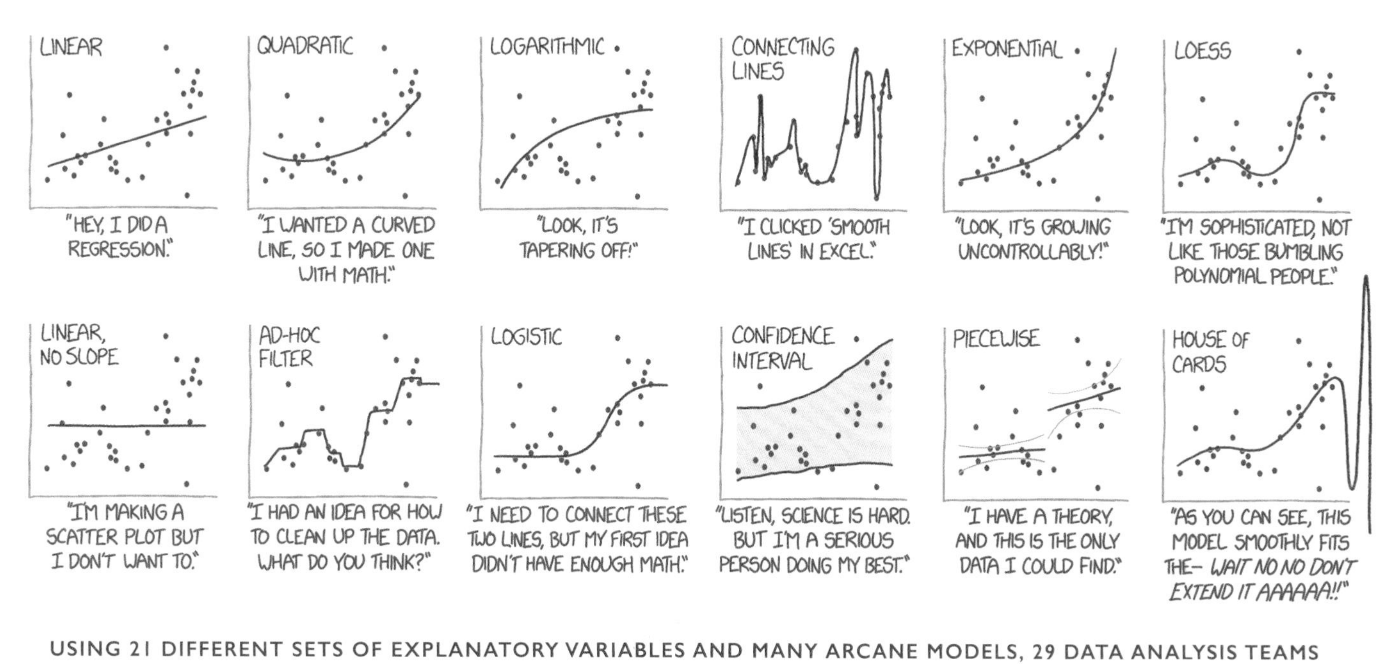
xkcd CURVE-FITTING METHODS AND THE MESSAGES THEY SEND
LINEAR
"HEY, I DID A REGRESSION."
QUADRATIC
"I WANTED A CURVED LINE, SO I MADE ONE WITH MATH."
LOGARITHMIC
"LOOK, IT'S TAPERING OFF!"
CONNECTING LINES
"I CLICKED 'SMOOTH LINES' IN EXCEL."
EXPONENTIAL
"LOOK, IT'S GROWING UNCONTROLLABLY!"
LOESS
"I'M SOPHISTICATED, NOT LIKE THOSE BUMBLING POLYNOMIAL PEOPLE."
LINEAR, NO SLOPE
"I'M MAKING A SCATTER PLOT BUT I DON'T WANT TO."
AD-HOC FILTER
"I HAD AN IDEA FOR HOW TO CLEAN UP THE DATA. WHAT DO YOU THINK?"
LOGISTIC
"I NEED TO CONNECT THESE TWO LINES, BUT MY FIRST IDEA DIDN'T HAVE ENOUGH MATH."
CONFIDENCE INTERVAL
"LISTEN, SCIENCE IS HARD. BUT I'M A SERIOUS PERSON DOING MY BEST."
PIECEWISE
"I HAVE A THEORY, AND THIS IS THE ONLY DATA I COULD FIND."
HOUSE OF CARDS
"AS YOU CAN SEE, THIS MODEL SMOOTHLY FITS THE– *WAIT NO NO DON'T EXTEND IT AAAAAA!!*"

USING 21 DIFFERENT SETS OF EXPLANATORY VARIABLES AND MANY ARCANE MODELS, 29 DATA ANALYSIS TEAMS PRODUCED MAYBE SOMEWHAT CONSISTENT FINDINGS FROM THE SAME SMALL DATA SET

STATISTICAL MODEL	ODDS RATIO	INTERVAL WIDTH
Zero-Inflated Poisson Regression	0.89	1.11
Bayesian Logistic	0.96	0.41
Hierarchical Log-Linear Modeling	1.02	0.03 ?
Multilevel Regression and Logistic Models	1.03	0.04 ?
Hierarchical Bayes Model	1.10	0.29
Logistic Regression	1.12	0.55
OLS Robust Standard Errors Logistic	1.18	0.46
Spearman Correlation	1.21	0.01 ?
WLS Regression with Clustered Standard Errors	1.21	0.49
Multiple Linear Regression	1.25	0.44
Clustered Robust Binomial Logistic Regression	1.28	0.53
Linear Probability Model	1.28	1.36
Hierarchical General Linear Model Poisson Sample	1.30	0.48
Multilevel Logistic Regression Bayesian Inference	1.31	0.48
Mixed-Model Logistic	1.31	0.46
Hierarchical Poisson Regression	1.32	0.57
Linear Probability Model, Logistic	1.34	0.53
Generalized Linear Mixed Models	1.38	0.65
Multilevel Logistic Regression	1.38	0.61
Mixed-Effects Logistic Regression	1.38	0.59
Generalized Linear Models for Binary Data	1.39	0.65
Negative Binomial Regression with a Log Link	1.39	0.48
Cross-Classified Multilevel Negative Binomial	1.40	0.56
Poisson Multilevel Modeling	1.41	0.62
Multilevel Logistic Binomial Regression	1.42	0.52
Generalized Linear Mixed-Effects Logit Link	1.48	0.64
Dirichlet-Process Bayesian Clustering	1.71	0.02 ?
Tobit Regression	2.88	10.44
Poisson Regression	2.93	78.55

ODDS RATIO: 0 1 2 3 4 5

"29 teams with 61 analysts used the same data set to address the same question: Are soccer referees more likely to give red cards to dark-skin-toned players than to light-skin-toned players? Analytical approaches varied widely across the teams, and the estimated effect sizes ranged from 0.89 to 2.93 (median = 1.31) in odds-ratio units. 20 teams (69%) found a statistically significant positive effect, and 9 teams (31%) did not observe a significant relationship. Overall, 29 different analyses used 21 unique combinations of covariates. Neither analysts' prior beliefs about the effect of interest nor their level of expertise readily explained the variation in the outcomes of analyses. Peer ratings of quality of analyses also did not account for variability. These findings suggest that significant variation in the results of analyses may be difficult to avoid, even by experts with honest intentions. Crowdsourcing data analysis, a strategy where numerous research teams are recruited to simultaneously investigate the same research question, makes transparent how defensible, yet subjective, analytic choices influence research results."

"Many Analysts, One Data Set," 46 co-authors and Brian Nosek, *Advances in Methods Practices in Psychological Science* 2018, edited

DATA MODELING: INHERENT ISSUES, BUT FEW KNOWN PREVALENCE RATES

⚠ MODELS WITH PLENTIFUL PARAMETERS CAN FIT PROTONS, ELEPHANTS, NOISE, WHATEVER

No matter how big one's proton detector, ever more extravagant Grand Unified Theory models can always be constructed that elude tests – such as symmetry groups E_6 or E_8, whose plentiful parameters can be tuned to make protons live as long as one pleases. One model might be correct, but no one would ever know. Dimitri Nanopoulos said: "People can construct models with higher symmetries and stand on their nose and try to avoid proton decay. OK, you can do it, but you cannot show it to your mother with a straight face." NATALIE WOLCHOVER

With four parameters I can fit an elephant, with five I can make it wiggle its trunk. JOHN VON NEUMANN

⚠ MODEL MULTIPLICITY AND CURVE-FITTING: WILLIAM FELLER'S ADVERSE REACTION

An unbelievably large literature tried to establish a transcendental "law of logistic growth." Lengthy tables, complete with chi-square tests, supported this thesis for human populations, bacterial colonies, development of railroads, etc. Both height and weight of plants and animals were found to follow the logistic law even though it is theoretically clear these 2 variables cannot be subject to the same distribution. height is linear, weight is volume *The trouble with the theory is that not only the logistic distribution, but also the normal, the Cauchy, and other distributions can be fitted to the same material with the same or better goodness of fit. In this competition logistic distributions play no distinguished role whatever; many theoretical models can be supported by the same observational material. Theories of this nature are short-lived because they open no new ways, and new confirmations of the same old thing soon grow boring. But the naive reasoning has not been superseded by common sense.* WILLIAM FELLER

⚠ PEOPLE CAN'T KEEP THEIR OWN SCORE: CLAIMING "MY DOG IS THE BEST DOG IN THE WORLD" DOES NOT VALIDATE YOUR MODEL

If someone shows you simulations that only show the superiority of their method, you should be suspicious. Good simulations will show where the method shines but also where it breaks. BYRAN SMUCKER & ROB TIBSHIRANI

Tuning your own method but insufficiently tuning the competing methods is one of those hidden problems in simulations for methods papers. MANJARI NARAYAN & ROB TIBSHIRANI

⚠ A GOOD MODEL EXPLAINS DATA, DOES NOT MEMORIZE DATA

A mesh to mimic content makes luscious animations, but mesh-drapery doesn't explain much. Overfitted models chasing data are brittle, breaking down and regressing toward the truth when confronted by new data. Computing millions of models is easy, but explaining something so well that it leads to replicated real-world explanations and successful interventions is very difficult.

In that Empire, the Art of Cartography attained such Perfection that the map of a single Province occupied the entirety of a City, and the map of the Empire, the entirety of a Province. In time, those Unconscionable Maps no longer satisfied, and the Cartographers Guilds struck a Map of the Empire whose size was that of the Empire, and which coincided point for point with it. The following Generations, who were not so fond of the Study of Cartography as their Forebears had been, saw that this vast Map was Useless, and not without some Pitilessness, they delivered it up to the Inclemencies of Sun and Winters. In the Deserts of the West, still today, there are Tattered Ruins of that Map; in all the Land there is no other Relic of the Disciplines of Geography. JORGE LUIS BORGES

⚠ FRESH DATA REMODELS MODELS: ON THE PHILLIPS CURVE, A FOUNDATIONAL MODEL IN MACROECONOMIC RESEARCH, TEXTBOOKS, AND POLICY-MAKING

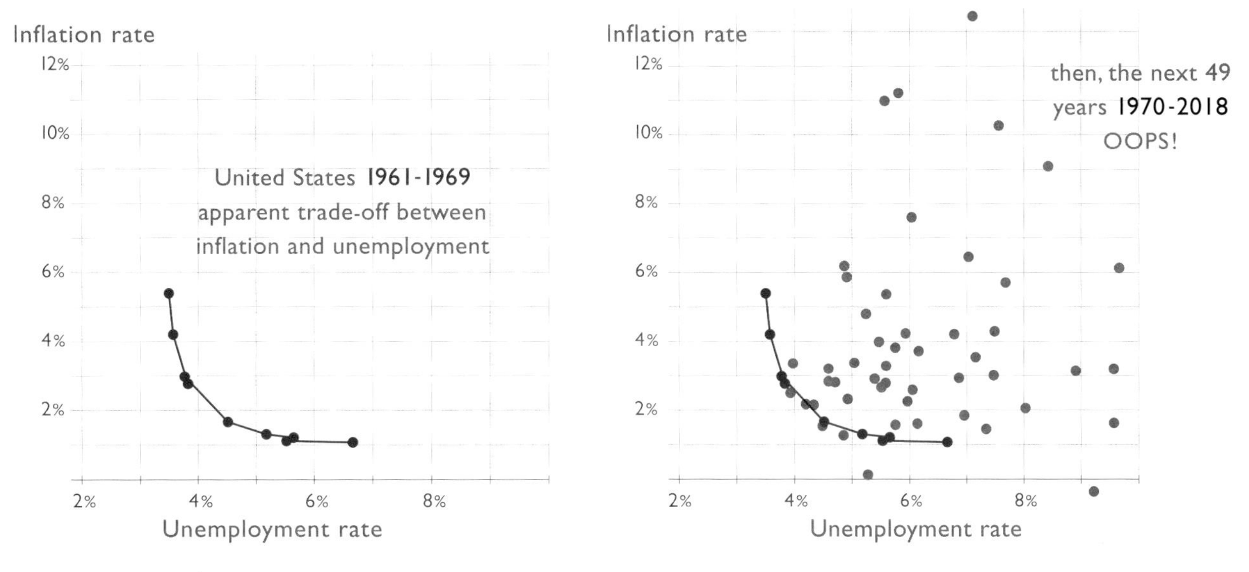

⚠ THE CURSES (AND BLESSINGS) OF HIGH-DIMENSIONAL DATA

"The curse of dimensionality arises when analyzing data in high dimensional spaces (often with hundreds or thousands of dimensions) that do not occur in low-dimensional settings such as the 3-dimensional physical space of everyday experience. The expression was coined by Richard Bellman when considering problems in dynamic optimization. Cursed phenomena occur in domains such as numerical analysis, sampling, combinatorics, machine learning, data mining, databases. When the dimensionality increases, the volume of the space increases so fast that the available data become sparse. In order to obtain a statistically sound and reliable result, the amount of data needed to support the result often grows exponentially with the dimensionality. Also, organizing and searching data often relies on detecting areas where objects form groups with similar properties; in high-dimensional data, however, all objects appear to be sparse and dissimilar in many ways." WIKIPEDIA

The curse is a mathematical truth, but it's more complicated than that. Advanced students may consult David Donoho, "High-Dimensional Data Analysis: The Curses and Blessings of Dimensionality," 2000.

⚠ SUBGROUP ANALYSIS: LEARNING FROM DATA VS. SUB-SUB-SUB-GROUP CHERRY-PICKING

In learning from data, subgroup analysis is essential. But, in workaday practice, a big problem: cherry-picking subgroups for desired findings, then publishing those findings as proven results, rather than as possibilities requiring replication. John Mandrola reported an extreme example of subgroup over-reach: "The clinical trial compared percutaneous transluminal coronary angioplasty (PTCA) against standard medical therapy for angina. The primary endpoint of death and myocardial infarction occurred in 6.3% of patients in the PTCA arm and only 3.3% in the medical arm. *But instead of saying angioplasty was twice as bad as medical therapy,* the abstract begins with this:

> In patients with coronary artery disease considered suitable for either PTCA or medical care, early intervention with PTCA [subgroup] was associated with greater symptomatic improvement [secondary endpoint subsubgroup], especially patients with more severe angina [secondary endpoint subsubsubgroup]."

MANY MEDICAL INTERVENTIONS ARE UNVALIDATED, INVALIDATED, LOW VALUE: IS NON-EVIDENCE BASED MEDICINE MORE PREVALENT THAN EVIDENCE-BASED MEDICINE?

"An analysis of 3,017 randomized controlled trials published in 3 leading medical journals *(Journal of the American Medical Association, The Lancet, New England Journal of Medicine)* identified 396 medical reversals. Low-value medical practices are either ineffective or cost more than other options but only offer similar effectiveness. Such practices can result in physical and emotional harms, undermine public trust in medicine, and have opportunity and financial costs. Identifying and eliminating low-value medical practices will reduce costs and improve care. Medical reversals are a subset of low-value medical practices and are defined as practices that have been found, through randomized controlled trials, to be no better than a prior or lesser standard of care."

Diana Herrera-Perez, Alyson Haslam, Tyler Crain, Jennifer Gill, Catherine Livingston, Victoria Kaestner, Michael Hayes, Dan Morgan, Adam S. Cifu, and Vinay Prasad, "A Comprehensive Review of 3000 Randomized Clinical Trials in 3 Leading Medical Journals Reveals 396 Medical Reversals," eLife. 2019; 8: e45183. published online 2019 June 11.

A short list of medical reversals

low dose aspirin for primary prevention of cardiovascular events
surgery for meniscal tear or knee arthritis (~1,000,000 surgeries yearly) magnesium supplements for leg cramps
vitamin pills to improve health multivitamins for prevention of cardiovascular disease
wearable tech for long term weight loss avoidance of peanut allergy by infant peanut exposure
tight blood sugar control in critically ill patients bedrest to prevent preterm birth
mammogram screening every 2 years mammogram screening for all women
MRI in breast cancer surgery compression stockings to reduce risk of deep vein thrombosis after stroke
intravenous drug administration during out-of-hospital cardiac arrest
epidural glucocorticoid injections for spinal stenosis (> $200 million annually)
carotid artery stenting (compared to surgery) for symptomatic carotid stenosis
Ginkgo biloba for preventing cognitive decline in older adults ($250 million annually)
vitamin E in primary prevention of cardiovascular disease
electrocardiographic/hemodynamic effects of diet supplements containing Ephedra
opioid-based analgesics for acute extremity pain in the emergency department
cardiovascular effects of intensive lifestyle intervention in type 2 diabetes
screening tests and all-cause mortality HRT for preventing chronic disease in post-menopausal women
corticosteroid treatment and intensive insulin therapy for septic shock in adults
screening tests and all-cause mortality rapid MRI for patients with low back pain
follow-up of blood-pressure lowering and glucose control in type 2 diabetes . . .

In 2013 a clinical evidence group reviewed data on the effectiveness of 3,000 National Health Service treatments:

11%	were rated beneficial
23%	likely beneficial
7%	trade-offs between benefits and harms
6%	unlikely beneficial
3%	likely ineffective or harmful
50%	unknown effectiveness !
100%	

Q.W. Smith, R.L. Street, R.J. Volk, M. Fordis, "Differing levels of clinical evidence," *Medical Care Research Review*, 2013, 70, 3-13

⚠ CONSTRUCTING PREDICTIVE MODELS IS EASY, AUTHENTICITY AND PRACTICAL USE ARE DIFFICULT

145 prediction models for covid-19 are "poorly reported, at high risk of bias, and their reported performance is probably optimistic. . . . The predictors identified could be considered as candidate predictors for new models. . . . Unreliable predictions could cause more harm than benefit in guiding clinical decisions." *Similar problems occur in ~70% of all medical testing, procedures, research.*

Laure Wynants, et al, 'Prediction models for diagnosis and prognosis of covid-19 infection: systematic review and critical appraisal,' BMJ 2020; updated April 5, 2020

Compounds found in carrots reverse Alzheimer's-like symptoms
IN MALE MICE
Researchers use gene editing with CRISPR to treat lethal lung diseases before birth
This gene could play a major role in reducing brain swelling after stroke
IN MALE MICE

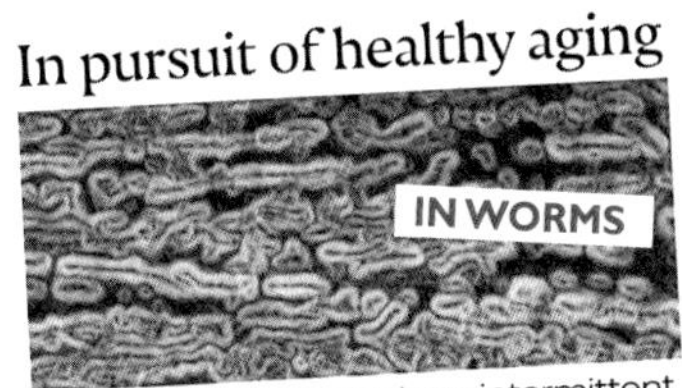

In pursuit of healthy aging
IN WORMS
Harvard study shows how intermittent fasting and manipulating mitochondrial networks may increase lifespan

⚠ "PROOF-OF-CONCEPT" AI MODELS ARE LIKE MOUSE STUDIES – EARLY RESULTS THAT CREATE FALSE HOPES FOR MEDICAL PATIENTS, MISLEADING PRESS RELEASES, PITCH SLIDES, PATENTS, COMMERCIAL EXPLOITATION. ALAS, RESEARCH AT THIS STAGE IS UNLIKELY TO EVER EXTEND LIVES.

On the difference between machine learning and AI:
If it is written in Python, it's probably machine learning.
If it is written in Powerpoint, it's probably AI. MAT VELLOSO

A 2019 study identified 516 recently published articles using AI algorithms for medical image diagnosis. Did these studies use TRIPOD standard methods recommended for clinical validation of AI performance: (1) external/internal validation, (2) diagnostic cohort/case-control research designs, (3) data from multiple institutions, (4) prospective research designs – methods recommended for clinical validation of AI performance? The conclusion: "Only 6% of the 516 studies showed proof-of-concept technical feasibility, necessary for validation of AI for clinical work." Also, some studies are compromised by leakage in high-dimensional spaces between model-building data and training data.

Dong Wook Kim, et al, 'Design Characteristics of Studies Reporting Performance of Artificial Intelligence Algorithms for Diagnostic Analysis of Medical Images,' *Korean Society of Radiology* 2019, edited.

BENEFITS OF VALIDATION, OPEN SOURCE, ETHICAL CHOICE OF PROBLEMS

Good research is rigorous and relevant, producing credible results that improve lives. Scientific credibility begins with honest independent quality assessments and replication with multiple data sets, especially for models with high-dimensional inputs. Half of all medical diagnostics and treatments are unvalidated, invalidated, or ineffective. These low-value practices are major targets of opportunity. Validated AI and machine learning are existential threats, and rightly so, to medical interventions with no better evidence than "we've always done it this way." AI/machine learning should avoid arcane proprietary models, closed data, secret sauces; avoid harming or bankrupting patients; and avoid the EHR business model. Such avoidances would provide competitive advantages over most existing diagnostics. Beneficial research should (1) not falsify, (2) have longer time-horizons than press releases, (3) turn replicated research into diagnostic *and* treatment protocols, (4) reduce medical care costs, (5) use open source code and open-published research articles.

	MARKED ENTHUSIASM	MODERATE ENTHUSIASM	NO ENTHUSIASM
RESULTS OF 6 WELL-DESIGNED (RANDOMIZED CONTROLLED) STUDIES:	0	3	3
RESULTS OF 47 POORLY DESIGNED STUDIES:	34	10	3

THOMAS CHALMERS, a founder of evidence-based medicine, demonstrated *the more susceptible a research design is to bias, the more enthusiastic the evidence for the favored treatments.* Replicated 1000s of times for medical interventions, this finding has survived for 50 years. For example, Chalmers and colleagues examined 53 published reports evaluating a surgery, portacaval shunts for esophageal bleeding. All reports were rated on (1) *enthusiasm* of findings favoring the surgery, (2) *quality of the research design* (good design = random assignment of patients to treatment or control; bad = treatment group not compared to any proper controls). The best design standard is the randomized controlled trial (RCT), which assigns patients randomly to the treatment or control group (assuring within chance limits that both groups are identical in all respects, known and unknown, and thereby avoiding, for example, selection of more promising patients to favored treatments). Of 53 published studies, only *6 were well-designed (RCT),* and none were markedly enthusiastic about the operation. In contrast, for *47 reports lacking valid controls,* 72% enthusiastically endorsed a procedure unwarranted by the RCT standard. (ET, *Beautiful Evidence,* 145, revised)

JOHN IOANNIDIS published his classic paper, "Why Most Published Research Findings Are False" in 2005. Research-on-research has since flourished, quantifying what the word "most" means – which ranges from 40% to 98% depending on medical specialties. Observational studies appear atop the leaderboard of untrue results. Randomized controlled double-blind studies are by far most likely to be true, as Thomas Chalmers decisively proved. Improving the credibility and integrity of research might extend millions of statistical lives by favoring better treatments and by avoiding unnecessary, ineffective, harmful, costly treatments.

PLOS | MEDICINE   OPEN ACCESS

75,143 saves   5,814 citations
2,698,716 views   7,704 shares
(July, 2021)

Why Most Published Research Findings Are False

John P. A. Ioannidis   August 30, 2005

There is increasing concern that most current published research findings are false. The probability that a research claim is true may depend on study power and bias, the number of other studies on the same question, and, importantly, the ratio of true to no relationships among the relationships probed in each scientific field. In this framework, a research finding is less likely to be true

when the studies conducted in a field are smaller
when effect sizes are smaller
when there is a greater number and lesser preselection of tested relationship
where there is greater flexibility in designs, definitions, outcomes, analytical modes
when there is greater financial and other interest and prejudice
when more teams are involved in a scientific field in chase of statistical significance.

Simulations show that for most study designs and settings, it is more likely for a research claim to be false than true. Moreover, claimed research findings may often be simply accurate measures of the prevailing bias. !

Meta-researchers do relevant and sometimes stunning replication studies: for example, work by Glenn Begley and Lee Ellis reports that only 11% of 63 landmark preclinical studies of cancer drugs could be reproduced, and by Brian Nosek and colleagues evaluating the "Replicability of Social Science Experiments in *Nature* and *Science*, 2010 to 2015."

Meta-research examines published journal articles – well-defined, stable, accessible populations. No missing data, no drop-outs, no randomization noise.

Meta-research is straight-forward, replicable, fast – (1) collect the research population, (2) identify bias, mess-ups, faults, falsification, validation issues, (3) replicate/confirm/adjudicate measurements (which, however, are often unblinded), (4) calculate prevalence of virtues and sins. Some measurements require judgments that might vary from judge to judge, such as 'spin' in abstracts, tactical citations (cites to other articles that are claimed to support the research findings, but in fact do not), whether subgroup analysis is worthy exploration or p-hacking. Precise definitions, independent replications, and weighted scoring may resolve such issues. Some measurements, however, are exact and replicable: specious accuracy, inappropriate image duplication, financial conflicts, binning, use of biomarkers not real outcomes, relative risks, and low-quality low-credibility research designs used in diet/nutrition studies, prediction modeling, oncology drug trials, and research on telepathy and flying saucers.

Meta-researchers assess credibility better than journal editors and their referees/reviewers. Speciality "reviews" are done by insider medical specialists, and usually fail to consider competing interventions beyond speciality fields, recognize new ideas, and rarely consider costs. Low-value research journals have published 1000s of sponsored, financially conflicted, biased, pseudo meta-research papers. Meta-researchers have a list of 800 ways to go wrong, sometimes with estimates of how much is wrong, how often it happens, when it matters. But empirical issues are turned into assumptions as in the slogan "unreliable evidence could cause more harm than benefits." How much harm? To whom? Under what conditions? What are the economic costs and opportunity costs of untrue studies? Thus meta-analysis teams should include health economists.

Conflicts of Interest A multiplicity of motives drives research: sincere beliefs that their idea will benefit the world – and also advance their careers via publications, consulting gigs, making big money for their sponsors and themselves. Meta-research has few financial conflicts because there is no money there.

Meta-research is often descriptive – identify a problem, assess its prevalence. But how can lousy/misleading research be quarantined, stopped, quickly identified? PubPeer and select Twitter groups have recently proved effective. Retractions take many years and lawyers.

The Grand Truths of Meta-Research

Well-designed randomized controlled trials are diamonds, many observational studies are quick sand.

Confirmation bias is omnipresent. *Money doesn't talk, it screams.* *It's more complicated than that.*

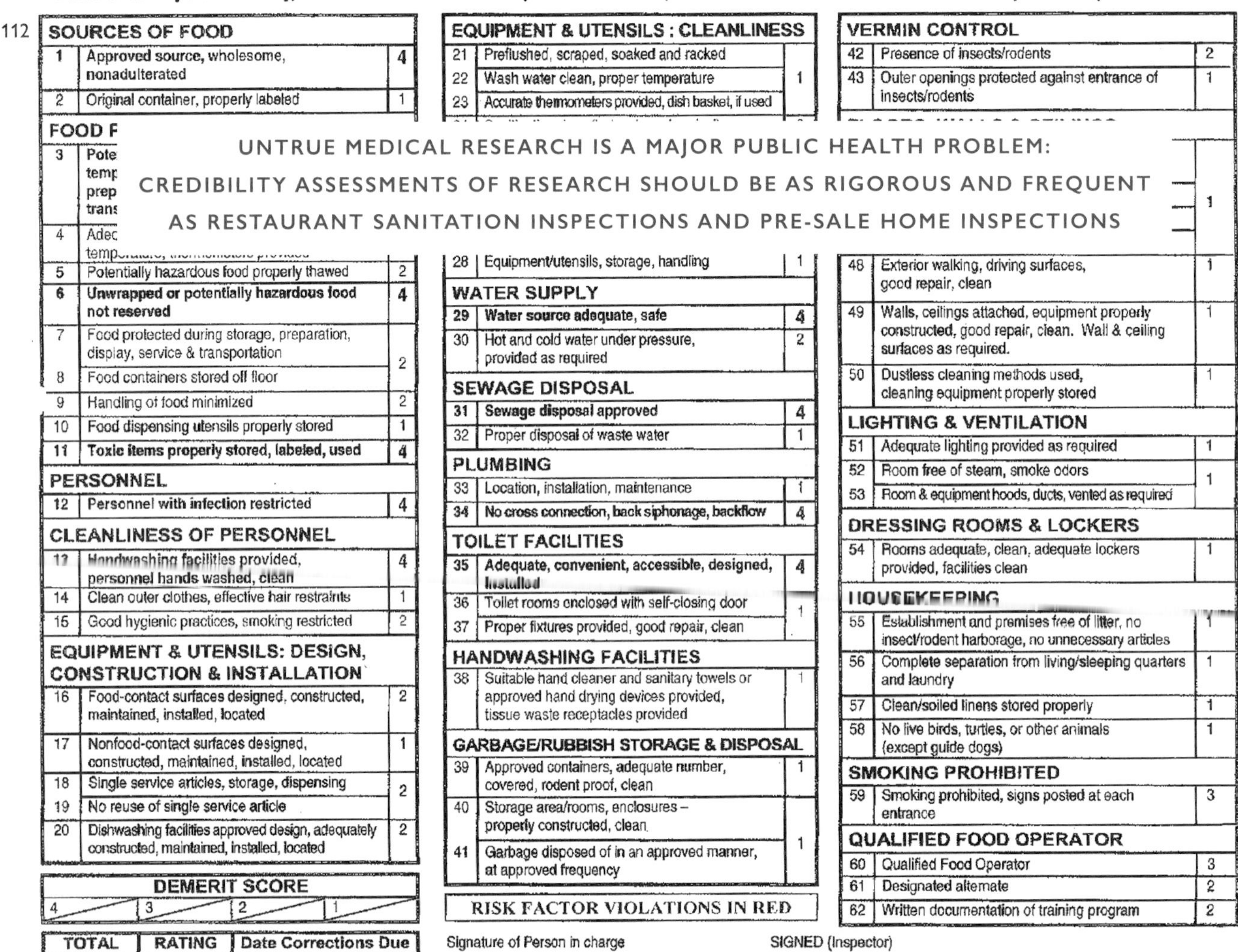

Based on an inspection this day, the items marked below identify the violations in operation or facilities which must be corrected by the date specified below.

SOURCES OF FOOD		
1	Approved source, wholesome, nonadulterated	4
2	Original container, properly labeled	1
FOOD P[illegible]		
3	Pote[illegible] temp[illegible] prep[illegible] trans[illegible]	
4	Adec[illegible] temp[illegible]	
5	Potentially hazardous food properly thawed	2
6	**Unwrapped or potentially hazardous food not reserved**	4
7	Food protected during storage, preparation, display, service & transportation	2
8	Food containers stored off floor	
9	Handling of food minimized	2
10	Food dispensing utensils properly stored	1
11	**Toxic items properly stored, labeled, used**	4
PERSONNEL		
12	Personnel with infection restricted	4
CLEANLINESS OF PERSONNEL		
13	Handwashing facilities provided, personnel hands washed, clean	4
14	Clean outer clothes, effective hair restraints	1
15	Good hygienic practices, smoking restricted	2
EQUIPMENT & UTENSILS: DESIGN, CONSTRUCTION & INSTALLATION		
16	Food-contact surfaces designed, constructed, maintained, installed, located	2
17	Nonfood-contact surfaces designed, constructed, maintained, installed, located	1
18	Single service articles, storage, dispensing	2
19	No reuse of single service article	
20	Dishwashing facilities approved design, adequately constructed, maintained, installed, located	2

EQUIPMENT & UTENSILS : CLEANLINESS		
21	Preflushed, scraped, soaked and racked	1
22	Wash water clean, proper temperature	
23	Accurate thermometers provided, dish basket, if used	
28	Equipment/utensils, storage, handling	1
WATER SUPPLY		
29	**Water source adequate, safe**	4
30	Hot and cold water under pressure, provided as required	2
SEWAGE DISPOSAL		
31	**Sewage disposal approved**	4
32	Proper disposal of waste water	1
PLUMBING		
33	Location, installation, maintenance	1
34	**No cross connection, back siphonage, backflow**	4
TOILET FACILITIES		
35	**Adequate, convenient, accessible, designed, installed**	4
36	Toilet rooms enclosed with self-closing door	1
37	Proper fixtures provided, good repair, clean	
HANDWASHING FACILITIES		
38	Suitable hand cleaner and sanitary towels or approved hand drying devices provided, tissue waste receptacles provided	1
GARBAGE/RUBBISH STORAGE & DISPOSAL		
39	Approved containers, adequate number, covered, rodent proof, clean	1
40	Storage area/rooms, enclosures – properly constructed, clean	1
41	Garbage disposed of in an approved manner, at approved frequency	

VERMIN CONTROL		
42	Presence of insects/rodents	2
43	Outer openings protected against entrance of insects/rodents	1
48	Exterior walking, driving surfaces, good repair, clean	1
49	Walls, ceilings attached, equipment properly constructed, good repair, clean. Wall & ceiling surfaces as required.	1
50	Dustless cleaning methods used, cleaning equipment properly stored	1
LIGHTING & VENTILATION		
51	Adequate lighting provided as required	1
52	Room free of steam, smoke odors	1
53	Room & equipment hoods, ducts, vented as required	
DRESSING ROOMS & LOCKERS		
54	Rooms adequate, clean, adequate lockers provided, facilities clean	1
HOUSEKEEPING		
55	Establishment and premises free of litter, no insect/rodent harborage, no unnecessary articles	1
56	Complete separation from living/sleeping quarters and laundry	1
57	Clean/soiled linens stored properly	1
58	No live birds, turtles, or other animals (except guide dogs)	1
SMOKING PROHIBITED		
59	Smoking prohibited, signs posted at each entrance	3
QUALIFIED FOOD OPERATOR		
60	Qualified Food Operator	3
61	Designated alternate	2
62	Written documentation of training program	2

DEMERIT SCORE: 4 / 3 / 2 / 1

TOTAL | RATING | Date Corrections Due

RISK FACTOR VIOLATIONS IN RED

Signature of Person in charge

SIGNED (Inspector)

UNTRUE MEDICAL RESEARCH IS A MAJOR PUBLIC HEALTH PROBLEM: CREDIBILITY ASSESSMENTS OF RESEARCH SHOULD BE AS RIGOROUS AND FREQUENT AS RESTAURANT SANITATION INSPECTIONS AND PRE-SALE HOME INSPECTIONS

Medical research reports, directly relevant medical care and clinical guidelines, must be *independently* inspected – as carefully and frequently as health/sanitation officials inspect restaurants, hairdressers, water supplies, sewage systems. Or, in selling a house, where both parties and a mortgage bank bring in independent inspectors. This 42-page pre-sale report shows 55 photos, including infrared images of electrical switch boxes. Or, how about fact-checking of medical research as done routinely in serious news reporting and some nonfiction publications?

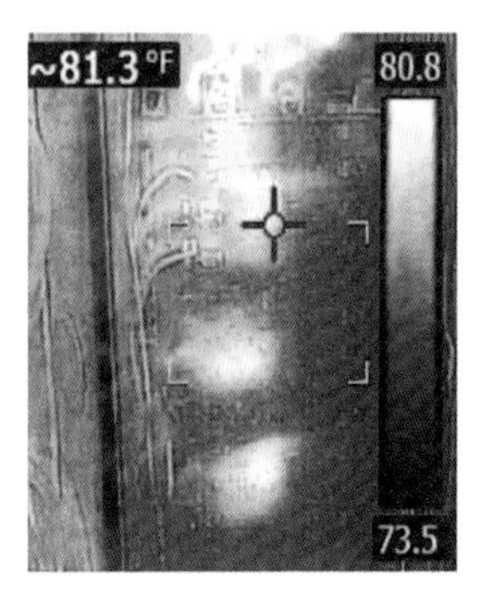

The Fact Checker's Bible

A GUIDE TO GETTING IT RIGHT

SARAH HARRISON SMITH

Performance audits and evaluations are always in danger of being distorted by financial interests, by compromised and corrupt bureaucracies and politicians, and by all of us who don't like arbitrary, fussy, inefficient, slow-motion bureaucracies telling us what to do. *But regulatory failure can turn into a massacre (270,000 U.S. deaths are attributed to oxycodone).* Every single oxycodone pill was approved by the U.S. Food and Drug Adminstration, and was made by licensed drug companies, prescribed by licensed doctors, sold by licensed pharmacists. All 72,000,000,000 pills (500 pills/U.S. household) were tracked to the exact place/time/amount of sale by the Drug Enforcement Agency. The only thing worse than regulatory agencies is the *Regulatory Theater of failed regulatory agencies captured, owned, and corrupted by those they regulate.*

QUICK CREDIBILITY SCORING OF RESEARCH ON HUMANS

substantive explanatory theory is vague, scientifically impoverished -7 randomized controlled trial +7
failure to report relative *and* absolute risk in the same paragraph -5 empirical assessment of measurement error +4
spurious correlation (eg, income drives both alleged cause and effect) -8 over-fitting -4
unconflicted funding of research +5 independent honest validation/replication +8
forensic audit of spreadsheet +5 undeclared financial conflict of interests -1 per $25,000 to each author
diet/nutrition study -5 contractor/sponsor/researcher have prior history of publication bias/retractions -7
inappropriate image duplication -5 failed randomization -6 binning -5 severe multicollinearity -5
summary models shown without underlying data -5
p-hacking/model-hacking/subgroup-hacking -7 midcourse changes in research protocol -3
ghostwritten or ghostgraphics -8 standard statistical model as sole assessment of error and uncertainty -6
use of unusual, magical, arcane statistical methods (pseudo-control groups, poorly chosen instrumental variables, inappropriate cross-over designs, etc) -5 results dependent on model assumptions -6
partied with sponsor's sales reps at professional meetings -0.5 each party, each author
and other high-prevalence pitfalls and biases . . . .

Scoring elements are based on universal principles of scientific inference. Thus epidemiologists, meta-researchers, and unconflicted biostatisticians should design scoring elements based on severity and prevalence. Fair scoring is global, neutral, indifferent to specialties and disciplines (no local, private, unscientific definitions of causality, patient outcomes, financial conflicts). To avoid inevitable attempts to game/evade/commercialize scoring systems, and to avoid the capture of scoring by those scored ('we're all on the same team'), an independent *Consumer Reports* model might be the best defense.

SCORING RESEARCH DESIGNS AND DATABASES

The same data from the same research design is often published multiple times; this may indicate highly productive research or salami-slicing or vanity over-publishing. Recall the case of compromised randomization in the Mediterranean diet study, with 267 published follow-up studies that relied on the same database – score one, score all 267. Database and research design credibility scores carry over each time the data are published (discounting for local over-fitting and sub-sub-group analysis, etc). This reduces scoring costs.

More than half of credibility scoring can be automated. Financially unconflicted biostatisticians, meta-researchers, and epidemiologists can do the rest. Medical specialists have already had their say in the reviewing and editing of journal articles of their fellow specialists. Scoring should be replicated independently; if scores diverge, a third or even fourth scoring can adjudicate.

Research credibility and integrity scores can be totaled up for all sorts of interesting clusters: high-impact articles, research teams, specialties, sub-specialties, sponsors, journals, paywalled versus open source, publishers, university research centers, laboratories, guidelines – and the articles, databases, and research designs referenced/reported in new drug approvals.

FROM RESEARCH TO DAILY PRACTICE: MEDICAL GUIDELINES HAVE SERIOUS CREDIBILITY ISSUES, AND MAY CAUSE/JUSTIFY LUCRATIVE FALSE ALARMS, OVER-DIAGNOSIS, OVER-TREATMENT.

People of the same trade seldom meet together, even for merriment and diversion, but the conversation ends in a conspiracy against the public, or in some contrivance to raise prices. ADAM SMITH

It is difficult to get people to understand something, when their salary depends on not understanding it. UPTON SINCLAIR

I've worked with 1000s of experts in psych diagnosis. Not one ever said 'Let's tighten criteria for my favorite diagnosis.' All worried about missed cases, none about harms and risks of mislabeling. ALAN CASSELS

It is simply no longer possible to believe much of the clinical research that is published, or to rely on the judgement of trusted physicians or authoritative medical guidelines. MARCIA ANGELL

Professional Societies Should Abstain From Authorship of Guidelines and Disease Definition Statements

John P.A. Ioannidis ***Circulation,*** **October 2018 edited**

"Guidelines from professional societies are increasingly influential. These documents shape how disease should be prevented and treated and what should come within the remit of medical care. Changes in definition of illness increases by millions the number of people who deserve specialist care: hypertension, diabetes mellitus, composite cardiovascular risk, depression, rheumatoid arthritis, or gastroesophageal reflux. Similarly, changes in prevention or treatment options may escalate overnight the required cost of care by billions of dollars. Should field specialists prepare such influential articles?

Professional society documents are written exclusively by insiders. Joining guideline panels is considered highly prestigious and allocation of writing positions is a unique means to advance an expert's visibility and career within the specific medical specialty. Tens of thousands of society members then cite these articles. Writing guidelines promotes the careers of specialists, building sustainable hierarchies of clan power, boosting the impact factors of specialty journals, elevating the visibility of the sponsoring organizations and conferences that massively promote society products to attendees. But do they improve medicine or do they promote biased, collective, organized ignorance?

Most published guidelines have red flags: sponsoring by a professional society with substantial industry funding, conflicts of interest for chairs and panel members, stacking, insufficient methodologist involvement, inadequate external review, and noninclusion of nonphysicians, patients, and community members. After the 2011 Institute of Medicine report, several societies changed the composition of their panels to avoid florid financial conflicts and preclude direct industry funding in guideline development. They have also included some methodologists. In recent guidelines, cardiovascular societies have tried to include more primary care physicians, nurses, patients on their panels. However, it is unclear such representatives can exert much influence when embedded within a dominant majority of vocal specialists. Moreover, stacking of panels with specialists who have overt preferences is more difficult to avoid.

Some professional societies are huge financial enterprises. Producers of medical guidelines and disease definitions tend to be the largest financial players, with cardiology the leading example. For example, the annual American Heart Association budget in the fiscal year 2016-2017 was $912 million, 20% of which came from corporate support. 77% of 60 million Euro annual income of the European Society of Cardiology comes from industry. Securing objectivity is difficult when industry-manufactured interventions also procure much of the specialty income. *Would a society therefore advise its members to change jobs, if evidence proved their medical services a waste?*

An overspecialized worldview is a major disadvantage in making sound recommendations. Specialists do not compare their merchandise against the merchandise of other healthcare providers. Specialists and societies compete for the same pie of healthcare resources.

Different countries vary on guidelines being entrusted to government or professional societies. In the United Kingdom, the National Institute for Health and Clinical Excellence is authorized by the government to consider both efficacy *and* cost. The US Preventive Services Task Force is convened by the Agency for Health Research and Quality, but most powerful guidelines are issued by professional societies; these place less attention on cost containment. With skyrocketing healthcare expenditures, largely cost-unconscious guidelines make little sense."

"An alternative approach: specialists should not assume any major role in guidelines pertaining to their own fields. Instead of having mostly or exclusively specialists write the guidelines and occasional nonspecialists consult or comment, guidelines could be written by methodologists and patients, with content experts consulted and invited to comment. This approach has been proposed also for systematic reviews and meta-analyses that synthesize the evidence feeding into guideline development. Another possibility is to recruit to the writing team other medical specialists who are unrelated to the subject matter. Involvement of such outsiders (for example, family physicians) could be refreshing. These people may have strong clinical expertise, but no reason to be biased in favor of the specialized practices under discussion. They may scrutinize comparatively what is proposed, with what supporting evidence, and at what cost. Devoid of personal stake, they can compare notes to determine if this makes sense versus what are typical trade-offs for evidence and decisions in their own, remote specialty. For example, while insider specialists may be willing to endorse an effective but highly expensive drug or device, outsiders may see more easily that this intervention is outrageously expensive. What may seem crucially important to a field expert, may appear as minutiae to a less personally involved outsider. Methodologists, patients, and different field specialists add to guideline teams more methodological rigor, patient-centeredness, and impartiality."

SPECIALTY GUIDELINES GOVERN YOUR MEDICAL CARE

Internal medicine doctor: "Esophagram shows C7 spinal osteophyte. Nothing to worry about."

Patient (ET): "Should a spine surgeon take a look?"

Internal medicine doctor: "No. He will operate on you."

Patient thinking: ["That's the best medical advice I have ever received."]

When you choose a specialty, you choose diagnosis, treatment – and specialty guidelines. If you think that's not a problem, read the definitive National Academy of Sciences Institute of Medicine report, *Clinical Practice Guidelines We Can Trust* or read the Ioannidis excerpt again.

RESISTANCE AND RESENTMENT TOWARD STATISTICAL EVIDENCE

"I do angioplasty and I have grateful patients. It's not rocket science for me to figure out if PCI [percutaneous coronary intervention] works or not," the chair of a leading cardiology department

Of course patients are grateful after an intervention – still alive, pain gone, high on anesthesia, told "you did great, everything went fine." Post-op patient gratitude, applause ending PCI theater, does not prove an intervention actually works or is worth the immense cost compared to untheatrical alternatives (exercise, diet, drugs). Self-congratulatory measurements made by those keeping their own score are contrary to the proven truth, replicated 1000s of times: *poor research designs create false, but desired, findings.* Huge gains in cardiac health have come from *randomized controlled trials* that validated life-extending statins and from big data demonstrating the enormous cardiac health benefits of smoking cessation. Do specialists reject this evidence for being insufficiently anecdotal?

High levels of short-term patient satisfaction appear to be associated with hospitality (greeters at the door, empathetic staff, comfortable rooms) – but also with *more treatments, higher costs, substantially higher mortality even after adjusting for baseline health and comorbidities.** Several plausible stories might explain these observational findings. But immediate post-treatment patient satisfaction/gratitude does not measure whether the patient lives better and longer, the ultimate goal. Can placebo effects be produced more cheaply and less dangerously than a $35,000 stent?

*Note titles of these reports: Joshua J. Fenton, et al, "The Cost of satisfaction: A national study of patient satisfaction, health care utilization, expenditures, mortality," JAMA Internal Medicine, 2012; and Cristobal Young and Xinxiang Chen, "Patients as consumers in the market for medicine: The halo effect of hospitality," Social Forces, 2020

'One day when I was a junior medical student, a very important Boston surgeon visited the school and delivered a great treatise on a large number of patients who had undergone successful operations for vascular reconstruction. At the end of the lecture, a young student at the back of the room timidly asked, "Do you have any controls?" Well, the great surgeon drew himself up to his full height, hit the desk, and said, "Do you mean did I not operate on half of the patients?" The hall grew very quiet then. The voice at the back of the room hesitantly replied, "Yes, that's what I had in mind." Then the visitor's fist really came down as he thundered, "Of course not. That would have doomed half of them to their death." God, it was quiet then, and one could scarcely hear the small voice ask, "Which half?"' E. E. PEACOCK

SELF-EVALUATION, CONTEMPT FOR DATA, FINANCIAL CONFLICTS

Investigative reporting by ProPublica/*The New York Times* discovered that financial conflict of interest statements made by doctors and researchers in their published articles diverged from their *actual* conflicts – serving on drug company Boards of Directors (a fiduciary, primary loyalty), creating private start-ups, receiving huge consulting payments. Some deans and researchers soon departed their positions, and 1000s of corrections were made to previously published research papers. The Memorial Sloan Kettering physician-in-chief resigned 3 days after the report came out. Then, in a *New York Times* interview, the MSK replacement physician-in-chief defended financial conflicts by first-person experiences:

> 'I'm telling you, as someone who works with patients, and I've worked with patients throughout my entire career here, that working with industry has helped me save lives. Maybe we should turn this around and say, we have more people on corporate boards because people value the opinions from our faculty.'

The Memorial Sloan Kettering chief of biostatistics/epidemiology had a different view:

> 'Bias can creep into the scientific enterprise in all sorts of ways. But financial conflicts are detectable definitively and represent a uniquely perverse influence on the search for scientific truth. The key substantive issue is that the problems we face were not caused by failures to disclose conflicts. The problems were due to the conflicts themselves. Making billions is not our mission. MSK is a nonprofit with a fundamentally social mission.'

In response, the MSK replacement physician-in-chief sneered at biostatisticians and data:

> 'He is a biostatistician. He lacks a full understanding of conflicts of interest. He does not work with patients. He works with data.'

WORKING WITH DATA, BIOSTATISTICIANS AND EPIDEMIOLOGISTS HAVE EXTENDED BILLIONS OF LIVES

Using vast amounts of data and intense detective casework in the field, epidemiologists measure new threats, then help respond, often successfully, to those threats. Their work, especially in vaccines and epidemics, has extended millions of lives and created billions of quality-life years. *Randomized controlled trials, designed and analyzed by biostatisticians, identify interventions that extend lives.* 'The most important medical advance in our generation is not a pill, or a stent, or a surgery, but the randomized controlled trial,' said Vinay Prasad. *Statistical analysis proved that smoking causes cancer*–leading to smoking cessation policies, preventing hundreds of millions of early deaths.

WALKING IN THE SHOES OF STATISTICAL LIVES:
40 YEARS OF CANCER DATA DESCRIBES MILLIONS OF INDIVIDUAL PATIENT EXPERIENCES,
EVALUATES CANCER SCREENING TESTS AND SUCCESS OF CLINICAL CARE

All cancers, rates of disease diagnosis ("you have cancer") and mortality, U.S., 1975-2015

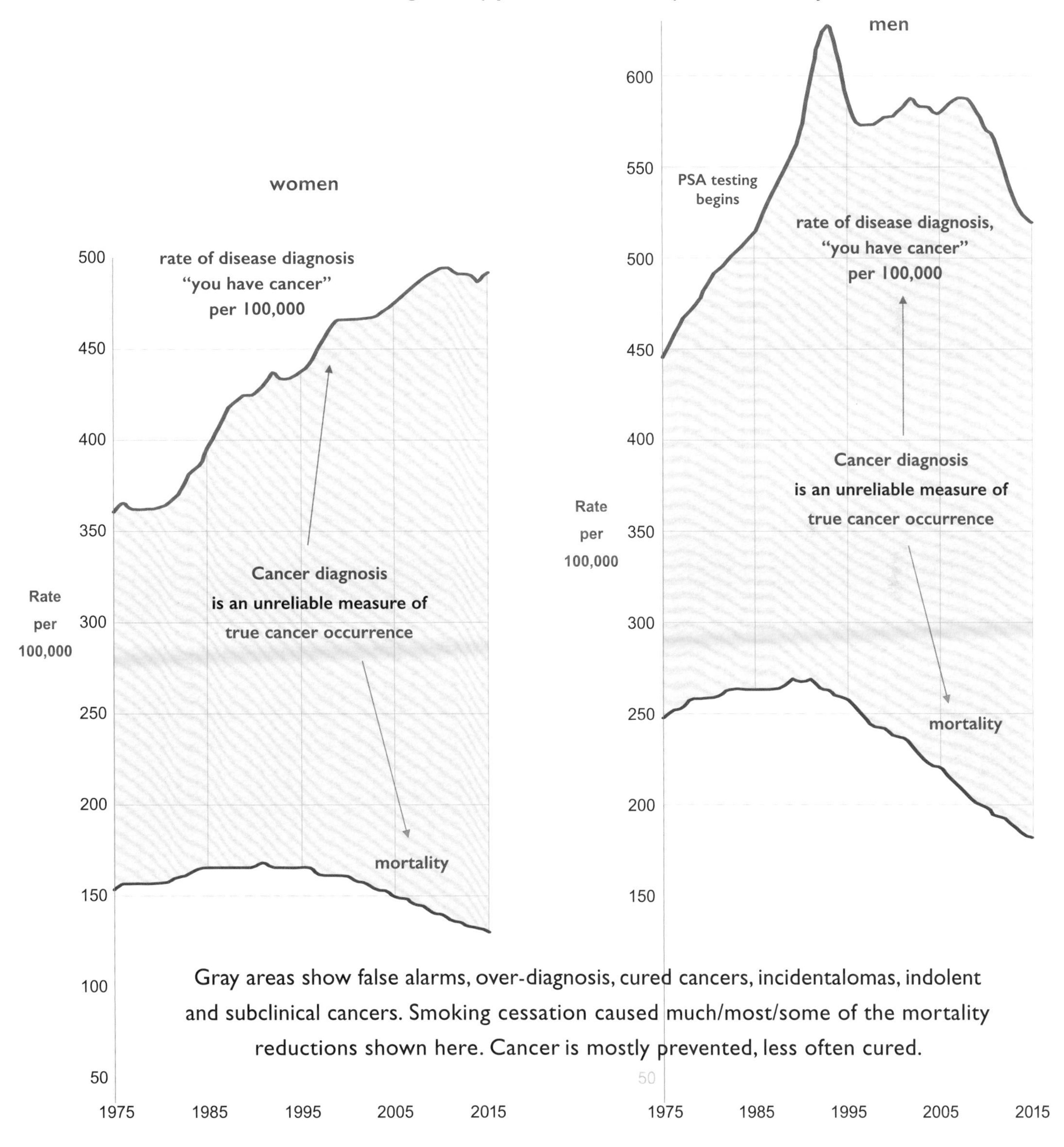

"Epidemiologic Signatures in Cancer," Gilbert Welch, Barnett S. Kramer, William C. Black, *NEJM*, October 2019, graphics redrawn. See also *Malignant: How Bad Policy and Bad Evidence Harm People with Cancer,* Vinayak K. Prasad, 2020.

KEEPING SCORE BY MEASURING STATISTICAL LIVES VALIDATES INDIVIDUAL VACCINATIONS, PREVENTS 103 MILLION DISEASE CASES

Building a data set of 88 million instances of disease each located in space and time from 1888 to 2011, epidemiologists were able to track the benefits of vaccination: a total of 103 million disease cases were prevented since 1924 in the United States. This natural experiment (before vs. after) shows strong consistent effects in all U.S. states for 5 diseases (measles, polio, rubella, hepatitis-A, mumps), as each state serves as its own control over the years – all adding up to direct visual proof.

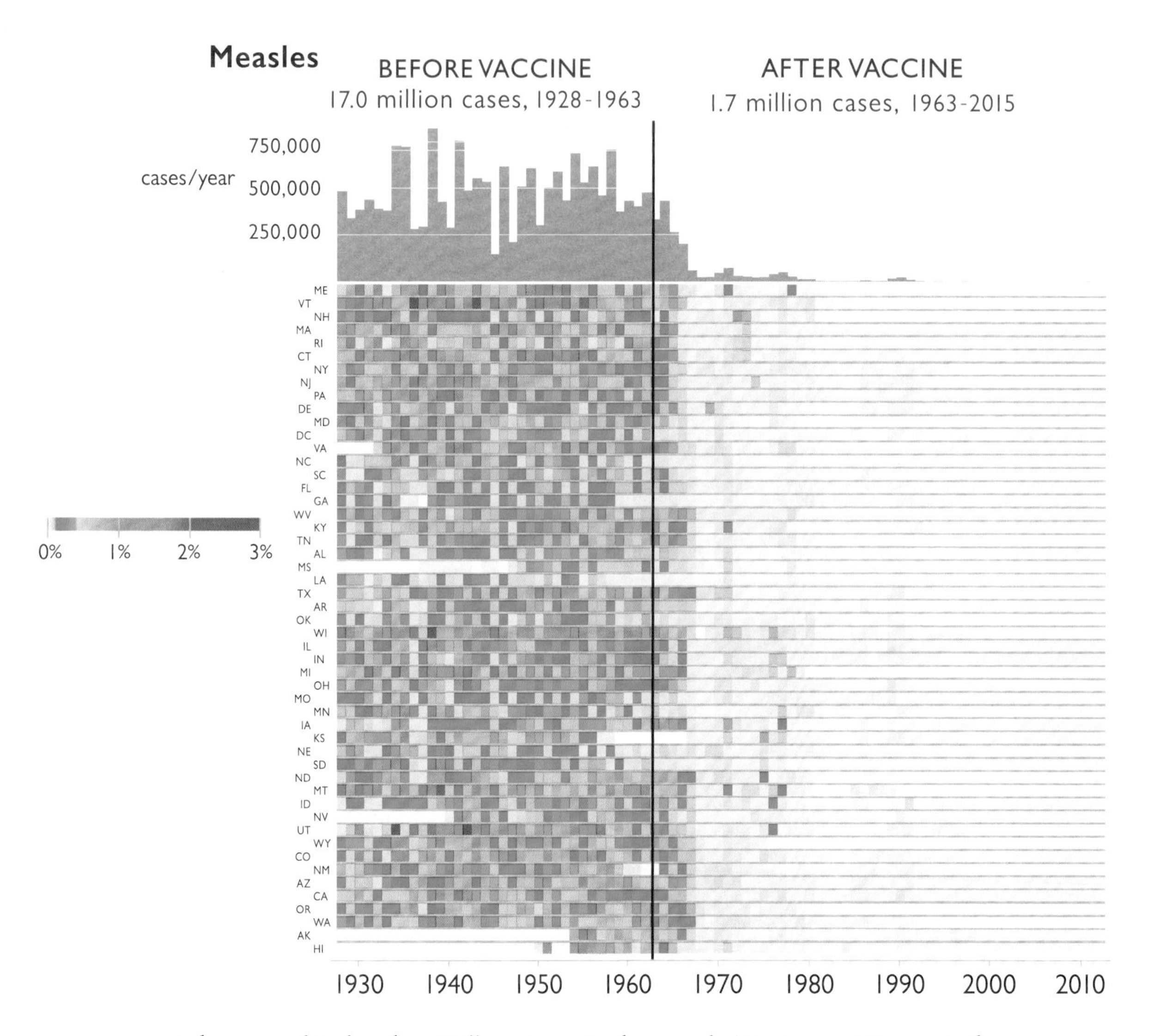

Redrawn graphics based on Willem G. van Panhuis, et al., 'Contagious Diseases in the United States from 1888 to the present,' *New England Journal of Medicine* 369, 2013, 2152-2158; and Tynan DeBold and Dov Friedman, 'Battling Infectious Diseases in the 20th Century: The Impact of Vaccines,' *The Wall Street Journal*, 11 February 2015, based on original data from the *Morbidity and Mortality Weekly Report*, Centers for Disease Control, compiled and analyzed by Project Tycho at the University of Pittsburgh.

~4,000,000,000 STATISTICAL LIVES: DIVERGENT PERFORMANCE

TRAJECTORIES IN A 2-DIMENSIONAL SPACE, 33 COUNTRIES, 1970 TO 2020

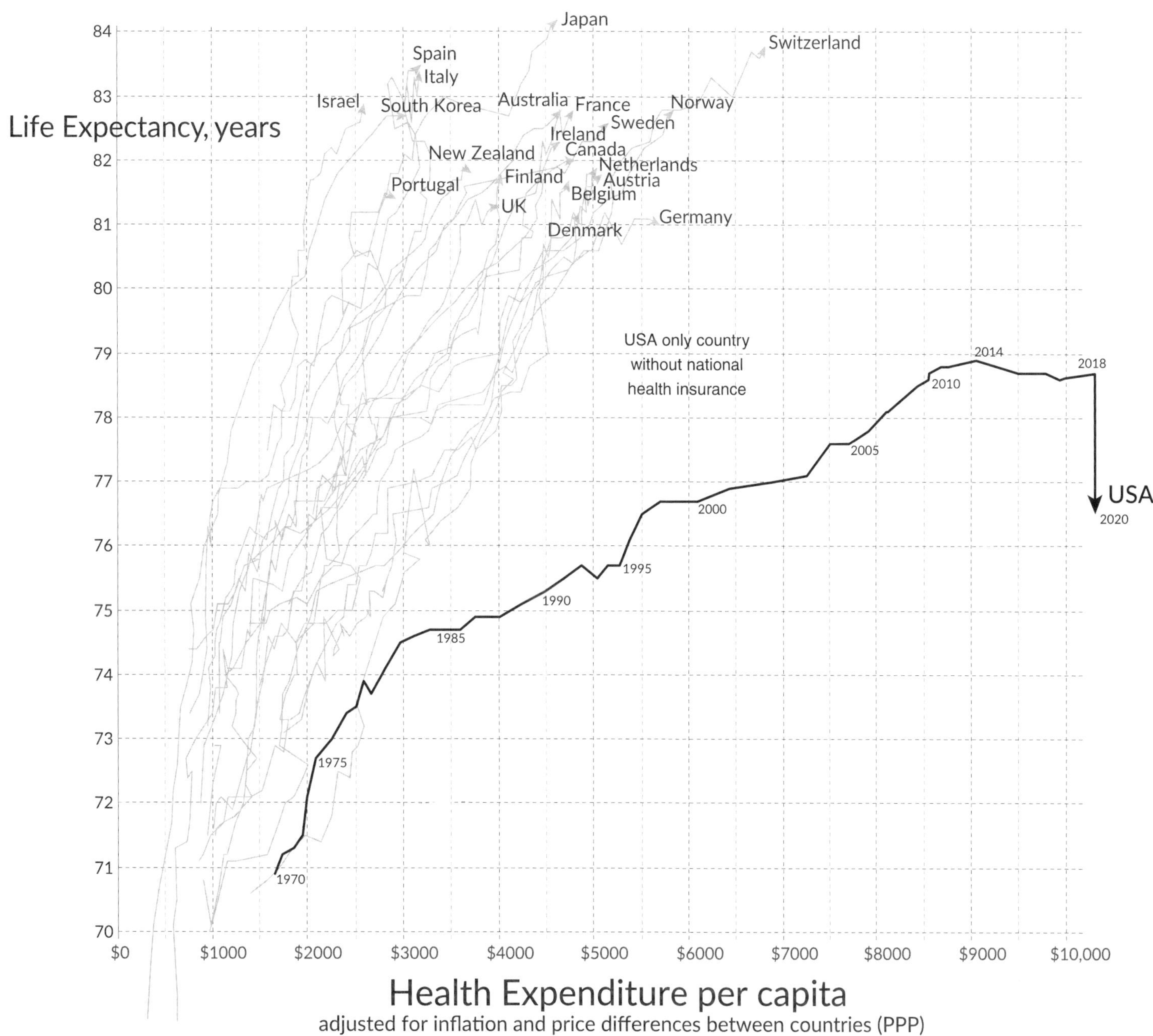

Data source: OECD — Note: Health spending measures the consumption of health care goods and services, including personal health care (curative care, rehabilitative care, long-term care, ancillary services, and medical goods) and collective services (prevention and public health services as well as health administration), but excluding spending on investments.
Shown is total health expenditure (financed by public and private sources).

OurWorldinData.org – Research and data to make progress against the world's largest problems.

The Metamorphosis

I

As Gregor Samsa awoke one morning from uneasy dreams he found himself transformed in his bed into a gigantic insect. He was lying on his hard, as it were armor-plated, back and when he lifted his head a little he could see his dome-like brown belly divided into stiff arched segments on top of which the bed quilt could hardly keep in position and was about to slide off completely. His numerous legs, which were pitifully thin compared to the rest of his bulk, waved helplessly before his eyes.

What has happened to me? he thought. It was no dream. His room, a regular human bedroom, only rather too small, lay quiet between the four familiar walls. Above the table on which a collection of cloth samples was unpacked and spread out—Samsa was a commercial traveler—hung the picture which he had recently cut out of an illustrated magazine and put into a pretty gilt frame. It showed a lady, with a fur cap on and a fur stole, *sitting upright and holding out to the* spectator a huge fur muff into which the whole of her forearm had vanished!

19

Vladimir Nabokov, in his teaching copy of Franz Kafka, *The Metamorphosis,* annotated the opening page with drawings of a gigantic beetle, and then argued with the English translation from original German. Kafka's text was substrate of Nabokov's markup, as Bach's score was substrate for violinist Yehudi Menuhin's markup of the score for Bach, *Sonata No. 2 for Solo Violin.*

5 ANNOTATIONS: EXPLANATORY WORDS, NUMBERS, GRAPHICS, IMAGES, ORGANIZED BY CONTENT-RESPONSIVE LOCAL GRIDS. IS THINKING JUST ANNOTATING THE WORLD?

Information displays should be annotated – combining words, images, graphics, and whatever it takes to describe and explain something. Annotation calls out and explains information and, at the same time, explains to viewers how to read data displays. Good annotation is like a knowledgeable expert/teacher at the viewer's side pointing and saying "Now see how this works with that, how this might explain that . . ."

Five bird-songs blowing in the wind annotate five bird images in Kircher's *Musurgia universalis* (1650) in this delightful and prankish diagram. Like an opera score, birds/birdsongs/words all play together. To represent the bird-song's chromatic syntax, musical notation combines with non-lexical vocables, sounds without wordlike meaning – dum dee dum, abracadabra, bee-bop-a-lula.

PICASSO'S *GUERNICA* ANNOTATED

READ AND SEE INTENSELY, YOU ARE IN THE PRESENCE OF A GREAT PAINTING,

BEAUTIFULLY EXPLAINED BY AD REINHARDT, PUBLISHED IN *P.M.* DAILY NEWSPAPER, JANUARY 5, 1947

HOW TO LOOK at a mural

Some words about the Picasso "Guernica" mural by Ad Reinhardt *

Almost ten years ago, a small, quiet ancient holy Basque town, full of refugees, became the target for the first "total" air-raid. It was wiped out in three and a half hours. Two days later the most famous living artist, the freest and most prolific painter of all time, started a mural that was to become the greatest anti-fascist work-of-art and the most impressive and monumental painting of the twentieth-century. The mural (12 x 26 feet) represented the Spanish Loyalist Government at the Paris World's Fair (1937), later toured London and America's seven largest cities, was seen by over a million people, raised over $10,000 here to save many Spanish lives, and may be seen now at the Museum of Modern Art.

The mural is not a picture-copy-imitation of a real scene you might see, or a simple poster or banal political-cartoon which you can easily understand (and forget) in a few minutes, but a design that diagrams our whole present dark age. It is a painting of pain and suffering. It symbolizes human destruction, cruelty and waste, not in a local spot but all over our one-world. It challenges our (yours, too) basic ways of living, thinking and looking.

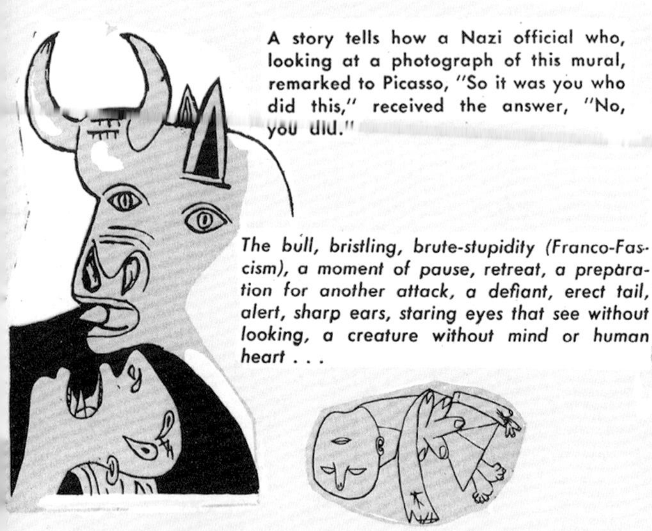

A story tells how a Nazi official who, looking at a photograph of this mural, remarked to Picasso, "So it was you who did this," received the answer, "No, you did."

The mural is an allegory. A pointing out of its symbolic-meanings will not explain its art-meanings (a last recombination of cubist and expressionist-surrealist-illustration, the end of a fine-art-picture-tradition). Here are, for what they are worth, some long labels. You see what you know and you don't look for what you don't want to see.

The design (photo-montage-like) is set in a self limited stage-world, an interior-exterior, an inside-room and an outside-town-square simultaneously. The color is black, white, pale and dark gray. (The dead have no color).

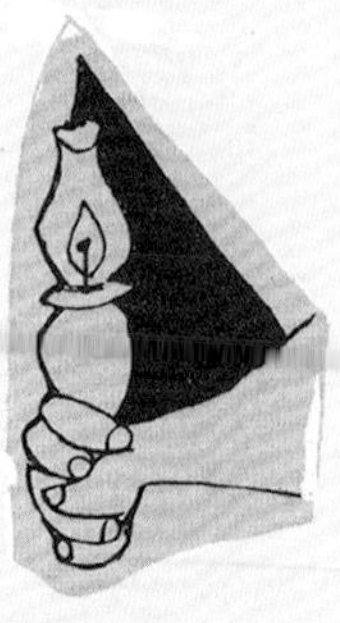

The bull, bristling, brute-stupidity (Franco-Fascism), a moment of pause, retreat, a preparation for another attack, a defiant, erect tail, alert, sharp ears, staring eyes that see without looking, a creature without mind or human heart . . .

A horse, dying, on one knee, (Spain), disemboweled, a stab in the back from above, a rigid tongue and raging throat, a twisted body, a newspaper-texture (cubist-pasted-paper) (headlines) . . .

A face aghast, a classic profile, a flying-thrust out of a window (like a fusion of fast photographic images), a terrified hold on breasts, an oil lamp (Truth) lights the tragedy.

A mother, anguished, a fierce cry to the skies, a pointed, paralyzed tongue, eyes and nostrils that become teardrops, a desperate pull up, a drag down, a lifeless child, limp as a rag-doll, lips sealed in innocence . . .

A sun (source of life), a radiant eye of the dark night, an electric-artificial-light-bulb (man's fateful discovery), all-seeing-God's-eye-witness.

A building aflame, an open-window, a falling, shrieking woman on fire, a half-opened door . . .

Ad Reinhardt, "How to look at a mural," published in *P.M.* daily newspaper, Brooklyn, NY, January 5, 1947

PICASSO'S *GUERNICA* ANNOTATED

READ AND SEE INTENSELY, YOU ARE IN THE PRESENCE OF A GREAT PAINTING, BEAUTIFULLY EXPLAINED BY AD REINHARDT, PUBLISHED IN *P.M.* NEWSPAPER, JANUARY 5, 1947

"No, painting is not done to decorate apartments. It is an instrument of war" . . . against "brutality and darkness"—Picasso

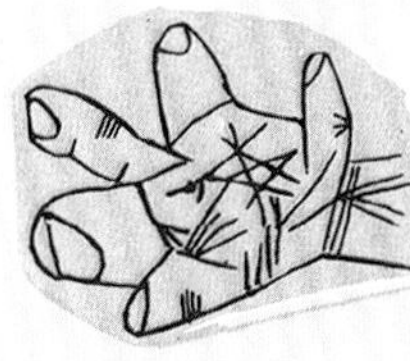

A hand, helpless, clumsy, chopped-off, scattered, divided fingers, stuck out like sore thumbs in all directions, a palm with crossed fatal life-lines . . .

A woman, dazed, fascinated, a looking-up-lunge, nipples bolted on breasts imprisoning maternity, futile appealing hands, a painful dragging of torn, broken feet . . .

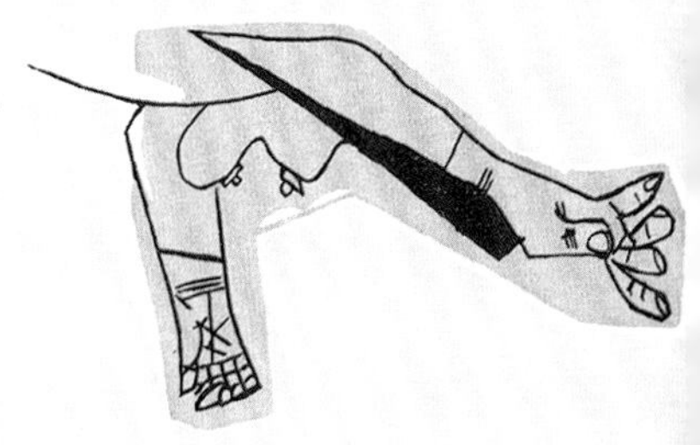

A man, decapitated, the bust of a smashed statue (the young Republic), eyes that cannot see straight, and roll in reflex-action, a look down into a hollow, open mouth with no sound, a look up under a nose (a circulating viewpoint from where you look), a cold, iron horse-shoe (good luck) turned as if seen from below, warns and threatens the onlooker (you) . . .

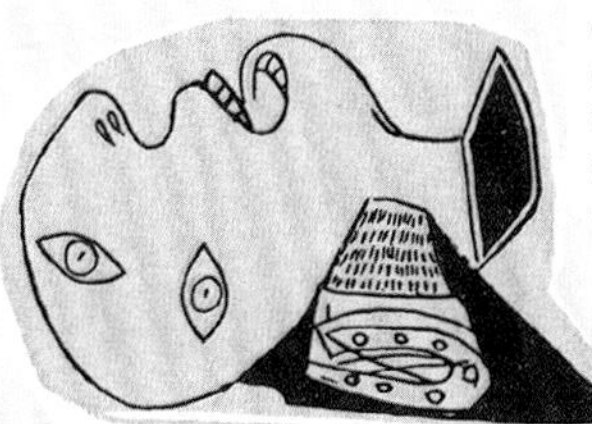

An arm, dismembered, mutilated, a clenched fist, a frozen grip of death, a broken sword, a hopeless defense gesture against a surprise attack, a young flower-blossom (renewal of life, hope) ". . . the last bud of the future"—Eluard.

Ad Reinhardt, "How to look at a mural," published in *P.M.* daily newspaper, Brooklyn, NY, January 5, 1947

MAKING A DATA REPORT BY ANNOTATING A DOCUMENT: AN EXPLANATORY CLASSIC

Macro-annotations explain micro-details of hospital costs in a 26-day narrative of a life and death in an intensive care unit. The design is transparent to the disturbing data, as a layered polyphony of voices — sequence, accounting data, commentary — weave together to trace out days, hours, minutes, money. * Note the extensive macro-data brought in by many annotations. These medical costs were incurred in 1982; to express in 2020 dollars, multiply by 8.

Revised from Edward Tufte, *Envisioning Information* (1990), 56 (redrawn from David Hellerstein, "The Slow, Costly Death of Mrs. K____," *Harper's*, 268 (March 1984), 84–89).

Mrs. K____ has been taken to the emergency room of a renowned hospital on Manhattan's Upper East Side. The doctors "work her up." More than $200 worth of blood tests are ordered ("emer rm lab," "lab serology out"), $232 worth of X-rays taken, $97.50 worth of drugs administered. I never saw Mrs. K____, she wasn't in my hospital, I don't know her medical history. But I am a doctor, and can reconstruct from her hospital bill what is going on, more or less. She is sick, very sick.

Mrs. K____ has been moved to the Intensive Care Unit ("room ICU"). It costs $500 a day to stay in the ICU, base rate. California has the highest average ICU rates in the country: $632 a day. In Mississippi, the average is $265. ICUs were developed in the 1960s. They provide technological life-support systems and allow for extraordinary patient monitoring. An inhalation blood-gas monitor ("inhal blood gas mont") is being used to keep a close check on the amount of oxygen in her blood. Without the attention she is receiving in the ICU, Mrs. K____ might already be dead.

Mrs. K____ has been running a high fever. The doctors have sent cultures of her blood, urine, and sputum to the lab to find out why. She is put on gentamicin ("lab gentamycin troug"), a powerful antibiotic. Such strong drugs can have toxic side effects. Gentamicin kills bacteria, but can also cause kidney failure.

It is Mrs. K____'s fifth day at the hospital, and she is slipping closer to death: her lungs begin to fail. She is put on a respirator ("inhal respirator"), which costs $119 a day to rent and requires a special technician to operate. A hospital can buy the machine for about $15,000.

BILL TO	INSURANCE COVERAGE
JOHN K____	BLUE CROSS STD 21

DATE	DESCRIPTION		TOTAL CHARGES
DETAIL	OF CURRENT CHARGES AND PAYMENTS		
09/23	EMER RM OTHER	5009000	119.00
09/23	EMER RM LAB	5006000	172.00
09/23	LAB SEROLOGY OUT	1406800	35.00
09/23	EMER RM EKG	5007000	61.00
09/23	X-RAY ABDOMEN	1501001	58.00
09/23	X-RAY CHEST RTN	1501009	58.00
09/23	X-RAY CHEST RTN	1501009	58.00
09/23	X-RAY CHEST RTN	1501009	58.00
09/23	PHARMACY	2601000	2.25
[illegible]	PHARMACY	[illegible]	[illegible]
09/23	PHARMACY	2601000	49.25
09/23	ROOM ICU		500.00
09/24	LAB AUTO BLOOD CT	1402101	17.00
09/24	LAB ACT PAR THROM	1404001	27.00
09/24	LAB PROTH DETER	1404011	17.00
09/24	LAB BLOOD CULT	1405002	37.00
09/24	LAB BLOOD CULT	1405002	37.00
09/24	LAB CHEM-20	1401104	31.00
09/24	SP HEM CBC	1602010	28.00
09/24	SP HEM RETIC CT	1602046	17.00
09/24	SP HEM PLATELET CT	1602090	17.00
09/24	LAB SEROLOGY OUT	1406800	35.00
09/24	LAB MAGNES	1401042	27.00
09/24	LAB RTN URINAL	1403001	16.00
09/24	LAB RTN CULT	1405003	37.00
09/24	LAB BACTERIA SM	1405011	16.00
09/24	LAB RTN CULT	1405003	37.00
09/24	LAB DIFF	1402099	15.00
09/24	LAB PROT ELEC	1401049	53.00
09/24	LAB FUNGUS	1405008	31.00
09/24	LAB FUNGUS	1405008	31.00
09/24	LAB TBC CULT	1405014	42.00
09/24	LAB DIFF	1402099	15.00
09/24	LAB AUTO BLOOD CT	1402101	17.00
09/24	X-RAY CHEST-BED	1501128	74.00
09/24	PHARMACY	2601000	10.00
09/24	PHARMACY	2601000	8.00
09/24	PHARMACY	2601000	8.00
09/24	PHARMACY	2601000	4.50
09/24	SPECIMEN MUCUS TRAP	2709085	3.00
09/24	SPECIMEN MUCUS TRAP	2709035	3.00
09/24	INHAL BLOOD GAS MONT	2101034	354.00
09/24	ROOM ICU		500.00
09/25	LAB SALICYLATE	1401050	49.00
09/25	LAB AMMONIA	1401006	40.00
09/25	LAB CHEM-20	1401104	31.00
09/25	LAB PROTH DETER	1404011	17.00
09/25	LAB CHEM-8	1401111	31.00
09/25	LAB BACTERIA SM	1405011	16.00
09/25	LAB AUTO BLOOD CT	1402101	17.00
09/25	LAB AUTO BLOOD CT	1402101	17.00
09/25	LAB ACT PAR THROM	1404001	27.00
09/25	LAB TBC CULT	1405014	42.00
09/25	LAB TBC CULT	1405014	42.00
09/25	LAB FUNGUS	1405008	3L.00
09/25	LAB RTN CULT	1405003	37.00
09/25	LAB RTN CULT	1405003	37.00
09/25	CARDIO ROUTINE EKG	1801001	61.00
09/25	X-RAY CHEST-BED	1501128	74.00
09/25	X-RAY ABDOMEN	1501001	58.00
09/25	X-RAY CHEST-BED	1501128	74.00
09/25	X-RAY CHEST-BED	1501128	74.00
09/25	PHARMACY	2601000	13.50
09/25	PHARMACY	2601000	39.00
09/25	PHARMACY	2601000	3.70
09/25	PHARMACY	2601000	16.50
09/25	PHAR IV SOLUTIONS	2601003	16.00
09/25	PHARMACY	2601000	2.50
09/25	PHAR IV SOLUTIONS	2601003	13.50
09/25	PHAR IV SOLUTIONS	2601003	13.50
09/25	PHARMACY	2601000	3.35
09/25	PHARMACY	2601000	2.25
09/25	INHAL BLOOD GAS MONT	2101014	354.00
09/25	ROOM ICU		500.00
09/26	LAB PROTH DETER	1404011	17.00
09/26	LAB CHEM-8	1401111	31.00
09/26	LAB AUTO BLOOD CT	1402101	L7.00
09/26	LAB UR SODIUM	1401077	27.00
09/26	LAB UR POTASS	1401076	27.00
09/26	LAB DIFF	1402099	15.00
09/26	LAB CHEM-8	1401111	31.00
09/26	LAB GENTAMICIN TROUG	1401112	27.00
09/26	CARDIO ROUTINE EKG	1801001	61.00
09/26	X-RAY CHEST BED	1501128	74.00
09/26	PHARMACY	2601000	31.20
09/26	PHARMACY	2601000	3.70
09/26	PHARMACY	2601000	13.50
09/26	PHARMACY	2601000	39.00
09/27	LAB CHEM-8	1401111	31.00
09/27	LAB PROTH DETER	1404011	17.00
09/27	LAB DIFF	1402099	15.00
09/27	LAB AUTO BLOOD CT	1402101	17.00
09/27	LAB ACT PAR THROM	1404001	27.00
09/27	LAB CHEM-8	1401111	31.00
09/27	LAB CHEM-20	1401104	31.00
09/27	LAB CHEMISTRY OUT	1401800	10.00
09/27	LAB FECES CULT	1405007	40.00
09/27	CARDIO ROUTINE EKG	1801001	61.00
09/27	BLD BK ANTIBDY SCRN	1701004	23.00
09/27	BLD BK ADMIN FEE	1701028	69.00
09/27	PHAR IV SOLUTIONS	2601003	37.50
09/27	PHAR IV SOLUTIONS	2601003	11.00
09/27	PHAR IV SOLUTIONS	2601003	26.00
09/27	PHAR IV SOLUTIONS	2601003	13.50
09/27	PHAR IV SOLUTIONS	2601003	63.50
09/27	PHARMACY	2601000	40.50
09/27	PHARMACY	2601000	11.00
09/27	PHARMACY	2601000	9.00
09/27	PHARMACY	2601000	13.50
09/27	PHARMACY	2601000	39.00
09/27	PHARMACY	2601000	3.70
09/27	PHARMACY	2601000	2.50
09/27	PACK CE 250 PROC FEE	1701018	46.00
09/27	25 NSA 50MU PROC FEE	1701077	35.00
09/27	INFUSION PUMP	2705027	30.00
09/27	INHAL RESPIRATOR	2102015	119.00
09/27	ROOM ICU		500.00
09/28	OPER OP RM 150	1001005	521.00
09/28	LAB OCC BLOOD	1403021	16.00
09/28	LAB GENTAMYCIN TROUG	1401117	27.00
09/28	LAB DIFF	1402079	15.00
09/28	LAB RTN CULT	1405003	37.00
09/28	CARDIO ROUTINE EKG	1801001	61.00
09/28	BLD BK GROUP RH	1701002	28.00
09/28	BLD BK X MATCH	1701006	46.00
09/28	BLD BK ANTIBDY SCRN	1701004	23.00
09/28	X-RAY CHEST-BED	1501128	74.00
09/28	X-RAY CHEST-BED	1501128	74.00
09/28	PHARMACY	2601000	11.00
09/28	PHAR IV SOLUTIONS	2601003	13.50
09/28	PHAR IV SOLUTIONS	2601003	50.00
09/28	PHARMACY	2601000	3.70
[illegible]	PHARMACY	[illegible]	[illegible]
09/28	PHARMACY	2601000	9.00
09/28	PHARMACY	2601000	13.50
09/28	ANEST ANEST DRUGS	1103001	12.40
09/28	INHAL RESPIRATOR	2102015	119.00
09/28	OPER OP RM SUPPLY	1002000	198.00
09/28	SUCT MACHINE-CONT	2704015	22.00
09/28	DIAL SOLN 1.5-CASE	2709040	24.00
09/28	INHAL BLOOD GAS MONT	2101034	354.00
09/29	LAB CHEM-8	1401111	31.00
09/29	LAB CHEM-20	1401104	31.00
09/29	LAB AUTO BLOOD CT	1402101	17.00
09/29	LAB DIFF	1402099	15.00
09/29	LAB DIAG SM/MILL	1407002	40.00
09/29	LAB AUTO BLOOD CT	1402101	17.00
09/29	LAB AUTO BLOOD CT	1402101	17.00
09/29	LAB DIFF	1402099	15.00
09/29	LAB DIFF	1402099	15.00
09/29	LAB GENTAMYCIN TROUG	1401112	27.00
09/29	LAB SP FL CELL CT	1402018	26.00
09/29	LAB CHEM-8	1401111	31.00
09/29	LAB CHEM-8	1401111	31.00
09/29	LAB CHEM-8	1401111	31.00
09/29	LAB FUNGUS	1405008	31.00
09/29	LAB BACTERIOLOGY OUT	1405800	35.00
09/29	LAB OVA & PARASITES	1405018	31.00
09/29	LAB SM&CELL BLOCK	1407003	53.00
09/29	LAB FIBRIN QUAN	1404007	40.00
09/29	LAB COAG FIBRIN SPLT	1404018	49.00
09/29	LAB ACT PAR THROM	1404001	27.00
09/29	LAB AUTO BLOOD CT	1402101	17.00
09/29	LAB FROZEN SECT	1408004	119.00
09/29	LAB RTN CULT	1405003	37.00
09/29	BLD BK COLD AGG	1701007	18.00
09/29	BLD BK ADMIN FEE	1701028	23.00
09/29	X-RAY CHEST-BED	1501128	74.00
09/29	X-RAY ABDOMEN	1501001	58.00
09/29	PHARMACY	2601000	13.50
09/29	PHARMACY	2601000	11.00
09/29	PHAR IV SOLUTIONS	2601003	50.00
09/29	PHARMACY	2601000	39.00
09/29	PHARMACY	2601000	3.70
09/29	ISOLATION GLOVES-BOX	2709025	7.00
09/29	HEEL-ELBOW PROTECTOR	2706025	9.00
09/29	HEEL-ELBOW PROTECTOR	2706025	9.00
09/29	DIAL-PERID KIT 87110	2708015	14.00
09/29	DIAL SOLN 1.5 CASE	2709040	24.00
09/29	SPECIMEN MUCUS TRAP	2709085	3.00
09/29	INHAL BLOOD GAS MONT	2101034	354.00
09/29	INHAL BLOOD GAS MONT	2101034	354.00
09/29	INHAL BLOOD GAS MONT	2101034	354.00
09/29	ROOM ICU		500.00
09/30	LAB AUTO BLOOD CT	1402101	17.00
09/30	LAB CHEM-8	1401111	31.00
09/30	LAB CHEM-8	1401111	31.00
09/30	LAB DIFF	1402099	15.00
09/30	SP HEM COAG STDY COM	1602007	239.00
09/30	SP HEMATOLOGY	1600000	49.00
09/30	SP HEM RETIC CT	1602046	17.00
09/30	SP HEM CBC	1602010	28.00
09/30	LAB BACTERIA SM	1405011	16.00
09/30	LAB ACT PAR THROM	1404001	27.00
09/30	LAB PROTH DETER	1404011	17.00
09/30	LAB FIBRIN QUAN	1404007	40.00
09/30	LAB AUTO BLOOD CT	1402101	17.00
09/30	LAB CHEM-20	1401104	31.00
09/30	LAB TBC CULT	1405014	42.00
09/30	LAB CHEM-20	1401104	31.00
09/30	LAB RTN CULT	1405003	37.00
09/30	LAB RTN CULT	1405003	37.00
09/30	BLD BK ADMIN FEE	1701028	207.00
09/30	X-RAY CHEST-BED	1501128	74.00
09/30	X-RAY CHEST-BED	1501128	74.00
09/30	PHAR IV SOLUTIONS	2601003	16.00
09/30	PHARMACY	2601000	39.00
09/30	PHAR IV SOLUTIONS	2601003	21.00
09/30	PHAR IV SOLUTIONS	2601003	16.00
09/30	PHARMACY	2601000	3.70
09/30	PHARMACY	2601000	13.50
09/30	PHARMACY	2601000	11.00
09/30	PHARMACY	2601000	2.25
09/30	PHAR IV SOLUTIONS	2601000	21.00
09/30	PHAR IV SOLUTIONS	2601000	21.00
09/30	PHAR IV SOLUTIONS	2601003	18.50
09/30	PHARMACY	2601000	2.50
09/30	PLAT CONC PROC FEE	1701014	180.00
09/30	FRSH FR PLA PROC FEE	1701019	26.00
09/30	INHAL RESPIRATOR	2102015	119.00
09/30	DRESSING SET-DISP.	2708041	7.00
09/30	VEST RESTRAINT	2709032	12.00
09/30	INHAL BLOOD GAS MONT	2101034	354.00
09/30	ROOM ICU		500.00
10/01	LAB CHEM-20	1401104	31.00
10/01	LAB CHEM-8	1401111	31.00
10/01	LAB CHEM-8	1401111	31.00
10/01	LAB CHEM-8	1401111	31.00
10/01	LAB DIFF	1402099	15.00
10/01	LAB AUTO BLOOD CT	1402101	17.00
10/01	BLD BK ADMIN FEE	1701028	23.00
10/01	X-RAY CHEST-BED	1501128	74.00
10/01	PHAR IV SOLUTIONS	2601003	37.50
10/01	PHARMACY	2601000	13.50
10/01	PHARMACY	2601000	11.00
10/01	PHARMACY	2601000	3.70
10/01	PHARMACY	2601000	31.20
10/01	PHARMACY	2601000	2.40
10/01	PHARMACY	2601000	27.20
10/01	PHAR IV SOLUTIONS	2601003	13.50
10/01	25 NSA 50MU PROC FEE	1701077	35.00
10/01	INHAL RESPIRATOR	2102015	119.00
10/02	PHARMACY	2601000	27.20

Mrs. K____'s first week in Intensive Care ends in a flourish of blood tests. She has five Chem-8s ("lab chem-8") — tests that measure the levels of sodium, potassium, and six other chemicals in her blood. The hospital charges Mrs. K____ $31 for each Chem-8. Most independent labs charge about half as much; some hospitals charge up to $60. The *New England Journal of Medicine* has said: "The clinical laboratory [is] a convenient profit center that can be used to support unrelated deficit-producing hospital operations." The *Annals of Internal Medicine* estimates that the number of clinical lab tests being done is rising 15 percent a year.

Mrs. K____ has started peritoneal dialysis ("dial-perid kit 87110"). Her kidneys are failing. She is still hooked up to the respirator. She is being kept alive by what Lewis Thomas calls "halfway technologies" — "halfway" because kidney dialysis machines and respirators can support organ systems for long periods of time, but can't cure the underlying disease. Some doctors are beginning to question this practice. A recent study at the George Washington University Medical Center concluded: "Substantial medical resources are now being used in aggressive but frequently futile attempts to avoid death."

Mrs. K____ has been put in a vest restraint. Restraints are used in Intensive Care to keep patients from thrashing about or pulling their tubes out. Many ICU patients develop what is called "ICU psychosis." They become disoriented, begin hallucinating. The condition is brought on by lack of sleep, toxic drugs, the noise of the ICU staff and machines, and pain.

FLOWING ANNOTATIONS OF FLOWING DATA

In 1610, Galileo reported the discovery of Jupiter's 4 moons in a 24 page narrative from *Sidereus Nuncius.* Nightly observations record locations of 4 moons over time – as words and images annotate each other. Reconstructed from Galileo's book, this stop-action stacklist below shows the first 11 days of seeing. On January 11, Galileo notes the moons *'wander around Jupiter, like Venus and Mercury around the Sun,'* endorsing Copernican heliocentrism:

Galileo, English translation	Latin original
Thus, on the twelfth, at the first hour of the following night, I saw the stars arranged in this manner. The more	OBSERVAT. SIDEREAE
East — West	Ori. — Occ.
eastern star was larger than the western one, but both were very conspicuous and bright. Both were two minutes distant from Jupiter. In the third hour a third little star, not at all seen earlier, also began to appear. This almost touched Jupiter on the eastern side and was very small. All were in the same straight line and aligned along the ecliptic. On the thirteenth, for the first time four little stars were seen by me in this formation with respect to Jupiter. Three were on the west and one on the east. They formed a very nearly straight line, but the middle star of the	Stella occidentaliori maior, ambæ tamen valdè conſpicuæ, ac ſplendidæ: vtra quæ diſtabat à Ioue ſcrupulis primis duobus; tertia quoque Stellula apparere cępit hora tertia prius minimè conſpecta, quæ ex parte orientali Iouem ferè tangebat, eratque admodum exigua. Omnes fuerunt in eadem recta, & ſecundum Eclypticæ longitudinem coordinatæ. Die decimatertia primum à me quatuor conſpectæ fuerunt Stellulæ in hac ad Iouem conſtitutione. Erant tres occidentales, & vna orientalis; lineam proximè
East — West	Ori. — Occ.
western ones was displaced a little to the north from the straight line. The more eastern one was 2 minutes distant from Jupiter; the intervals between the remaining ones and Jupiter were only 1 minute. All these stars displayed the same size, and although small they were nevertheless very brilliant and much brighter than fixed stars of the same size. On the fourteenth, the weather was cloudy. On the fifteenth, in the third hour of the night, the four stars were positioned with respect to Jupiter as shown	rectam conſtituebant; media enim occidẽtalium paululum à recta Septentrionem verſus deflectebat. Aberat orientalior à Ioue minuta duo: reliquarum, & Iouis intercapedines erant ſingulæ vnius tantum minuti. Stellæ omnes eandem præ ſe ferebant magnitudinem; ac licet exiguam, lucidiſſimæ tamen erant, ac fixis eiuſdem magnitudinis longe ſplendidiores. Die decimaquarta nubiloſa fuit tempeſtas. Die decimaquinta, hora noctis tertia in proximè depicta fuerunt habitudine quatuor Stellæ ad Iouem;
East — West	Ori. — Occ.
in the next figure. They were all to the west and arranged very nearly in a straight line, except that the third one from Jupiter was raised a little bit to the north.	occidentales omnes: ac in eadem proxim recta linea diſpoſitæ; quæ enim tertia à Ioue numerabatur paululum

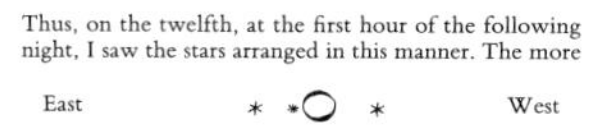

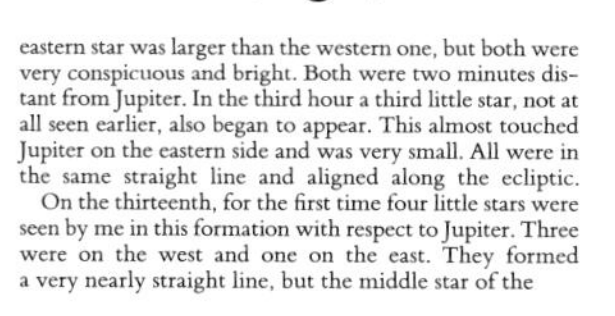

Day		
7		*With my new telescope, I saw 3 stars around Jupiter in a straight line parallel to the ecliptic.*
8		*Provoked by curiosity, tonight I viewed Jupiter again; now 3 stars were all west of Jupiter.*
9	*cloudy*	*Unfortunately tonight the sky was covered everywhere with clouds.*
10		*Only 2 stars, to the east! Is a third hiding behind Jupiter? Jupiter can't move that fast; thus the stars must move.*
11		*It appears that 3 stars wander around Jupiter, like Venus and Mercury around the Sun.*
12		*As I watched, a third star appeared nearest to Jupiter. All 3 little stars were straight on the ecliptic.*
13		*For the first time I saw 4 separate stars near Jupiter. The eastern star was 2 minutes from Jupiter,*
14	*cloudy*	*and the 3 other stars apart at about 1 minute intervals. But now, on the 14th, the weather was cloudy.*
15		*About 3 hours after sunset, 4 stars appeared; at 7 hours after, only 3 were in the same straight line.*
16		*Then 2 stars appeared within a distance of 40 seconds of Jupiter with 1 star 8 minutes from J in the west.*
17		*Again 2 stars; 4 hours later another emerged, which, I suspect, had been united with the first one.*

ANNOTATING NOVELS PAGE-BY-PAGE: JANE AUSTEN, PRIDE AND PREJUDICE

In this comprehensive annotated edition of *Pride and Prejudice,* every text page (left) is annotated on the adjacent page (right). The cover introduces this format, a design serving both avid and casual readers:

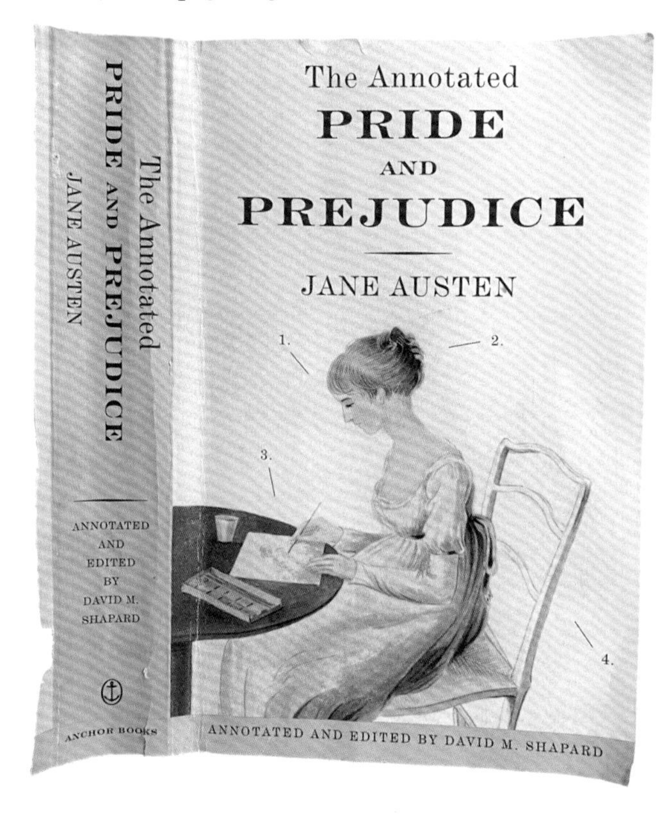

Annotations to the Front Cover

1 Jane Austen's sister, Cassandra, painted this picture of their niece, Fanny Austen Knight. Fanny's surname resulted from her father's adoption by distant relations named Knight who left their estate to him. This reasonably common procedure could explain the difference in name of Mr. Bennet and Mr. Collins in the novel, despite their common paternal ancestry.

2 Fanny was very close to Austen, often asking her advice about love and marriage. In her replies, Austen acknowledges the economic benefits of marriage for women but also argues firmly, in words she repeats almost verbatim in *Pride and Prejudice,* that one should never marry without love.

3 Fanny's sketching, like Cassandra's execution of this picture, suggests how many ladies drew or painted at that time. Such accomplishments were highly valued in young ladies; the Bennet girls are criticized for their inability to draw.

4 Fanny's dress is typical of period fashions, which favored high waists, soft flowing fabrics, and light colors.

MARK-UP ANNOTATIONS EVERYWHERE: LEONARDO DA VINCI'S ANATOMICAL DRAWINGS

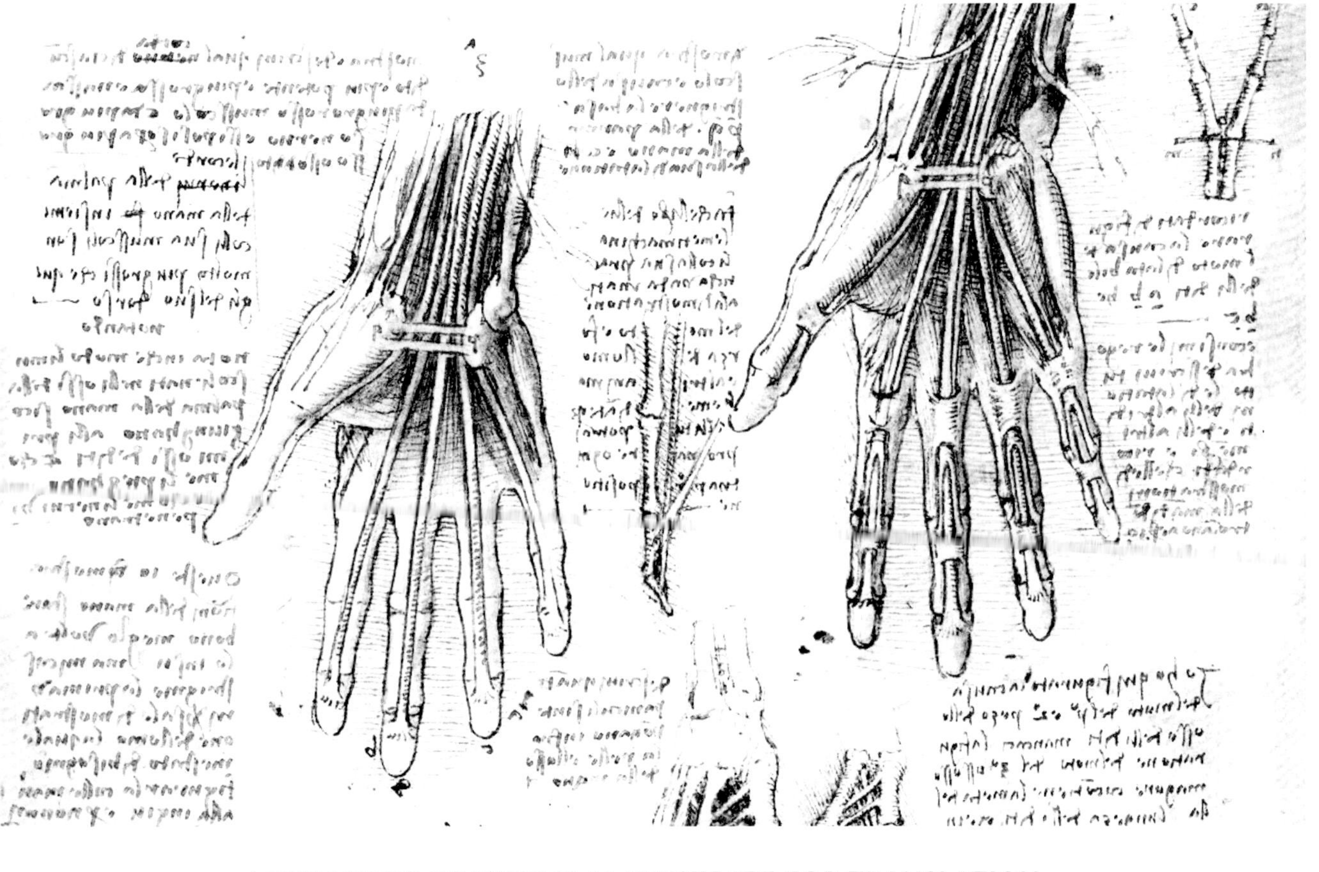

LEONARDO'S DRAWINGS AS SUBSTRATE FOR TRANSLATION

Show and describe what cord in each finger is the most powerful and the largest, and arised from the largest muscle and the largest sinew and is placed upon the largest digital bone.

The cords of the palm of the hand together with their muscles are very much larger than those of its dorsum.

to be noted

Note how the muscles [interossei] which arise from the bones of the palm of the hand are attached to the first bone of the digits, how they flex them and how the nerves are distributed there.

These 10 demonstrations of the hand would be better turned upwards, but my first general demonstration of man obliges me to do otherwise having had to

deep dissection of the palm of the hand

Demonstrate what muscle is the cause of the contraction of the base *p q*, of the palm of the hand and likewise, of its separation.

Arrange it so that the book on the elements of mechanics, with its practice, comes prior to the demonstration of movement and force of man and of other animals, and by means of these [examples] you will be able to prove all your propositions.

digital bones illustrating the insertion of the long flexor tendons

Describe how many coverings [panniculi] intervene between the skin and the

intermediate dissection of the palm of the hand

digital bones to illustrate abduction and adduction of the fingers

Remember to represent the cause of the movement of separation of the digits *a b* and *bc*.

And by the same rule, describe the separation of all the other digits and other members; and remember that the demonstrations of the interior of the hand must be 10.

I have represented here the cause of the motion of the first and 2nd segments of the digital bones; there remains the representation of the motion of the larger 3rd bone which

10 TEXT-PARAGRAPHS AND 10 MOTION-DIAGRAMS, ANNOTATING ONE ANOTHER. ANNOTATIONS ARE DIFFICULT TO UNSEE.

Diagrams and words depict/describe movement in 10 multimodal paragraphs. Constructed by the great news-graphics designer Megan Jaegerman, this became widespread refrigerator door art (now difficult in a digital world), living on beyond the daily news.

Muscle by Muscle: A Complete Strength-Training Workout

SHOULDERS
Lateral raises
Feet shoulder-width apart, knees slightly bent. Hold weights at your sides, slowly lift them outward from thigh level to shoulder level, elbows slightly bent. Slowly lower weights. Repeat.

SHINS
Toe raises
Sit on a stool or desk. With a light weight on your foot, slowly raise and lower the ball of the foot. Switch legs and repeat.

MIDSECTION
Curl downs
Sit with your knees bent, feet flat and arms reaching forward. Slowly lower yourself to the floor. Sit up again, using your arms if necessary. Repeat.

CALVES
Heel raises and dips
Stand with balls of your feet on a thick book or step. Slowly rise on your toes, then lower your heels as far as you can. Repeat.

BICEPS
Curls
Sit with your legs apart and one hand on your thigh. With a weight in the other hand, forearm horizontal and elbow on thight, curl the weight up toward your chest. Lower and repeat. Switch arms.

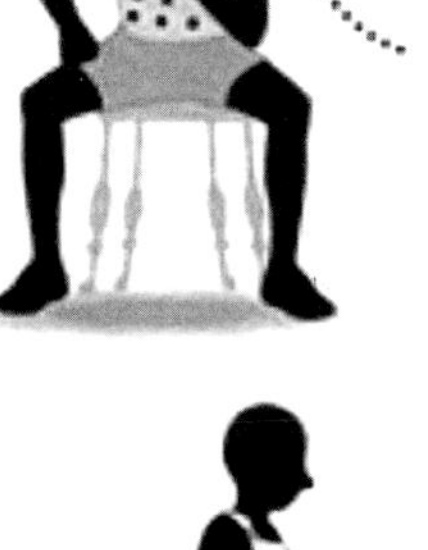

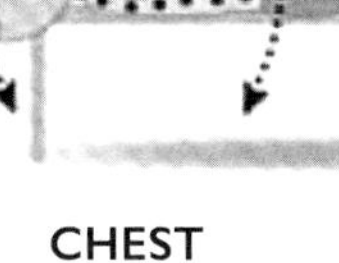

QUADRICEPS
Leg extensions
Sit on a stool or desk; put light weights on your ankles. Slowly straighten one leg, keeping your back straight and foot flexed.

CHEST
Bench fly
Lie on a bench; hold weights up over your chest. Slowly lower your arms in an outward arc until weights are at chest level. Reverse the movement, bringing weights back up. Repeat.

UPPER ARMS
Triceps extensions
With one knee and hand on a chair, hold a weight beside your chest, bending your arm at the elbow. Straighten your arm behind you; return to starting position. Switch arms and repeat.

HAMSTRINGS
Curls
Attach a light weight to your ankle, and hold on to a chair for support. Slowly lift your heel toward your buttocks, then lower it. Switch legs and repeat.

FOREARMS
Wrist curls
Holding a weight, rest your forearm on a table. With your hand over the edge, curl the weight up; then lower it as far as possible. Repeat.

Sources: Megan Jaegerman 1990-1998 portfolio, *The New York Times*. adapted from *The Wellness Letter*. Edited, reconstructed.

Country : Iceland

City : Búðardalur

Name : a horse farm with an icelandic / danish couple and 3 kids and a lot of sheep!

ÍSLAND INNANLANDS 65°08.73' 22°48.19' ELLIÐAEYJARVITI 50g

here

Laugar

Ísafjörður

Staðarfell

590

Hvammsfjörður

Búðardalur

60

Borgarnes

the danish woman works in a supermarket in Búðardalur.

Takk fyrir!

English/Icelandic postal delivery instructions (successful)   Iceland 2016

LEARN TO DRAW IN 4 EASY STEPS !

Sequential instructions for drawing, and behold . . . a miracle occurs!

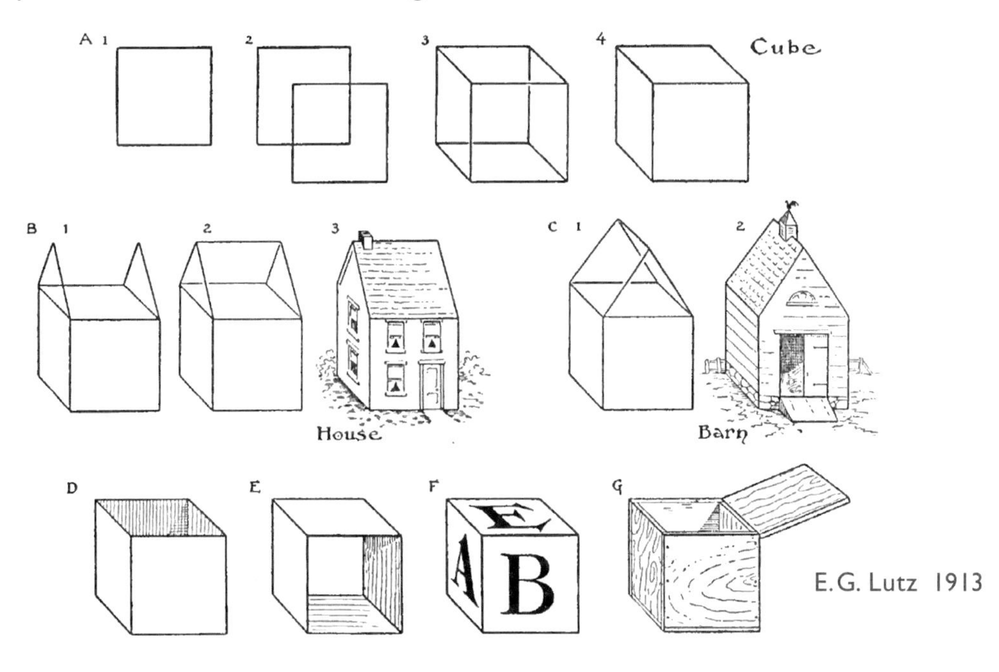

E.G. Lutz 1913

Direct instructions at point of need may encourage writers and programmers to divert diversions. Or not, because such signs are seen only a few times before becoming unseen.

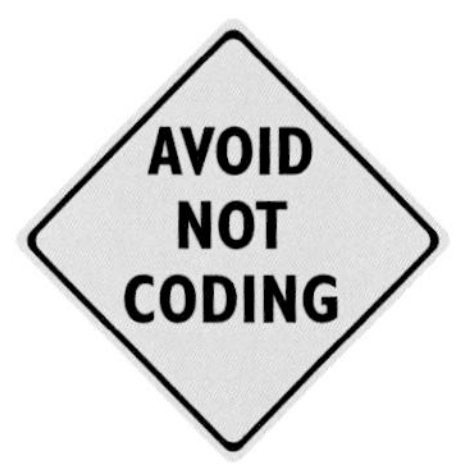

6 INSTRUCTIONS AT POINT OF NEED

Steps on Broadway Jack Mackie Seattle 1982

Nagasaki Prefectural Art Museum
Kenya Hara Yoshiaki Irobe
Hara Design Institute Japan

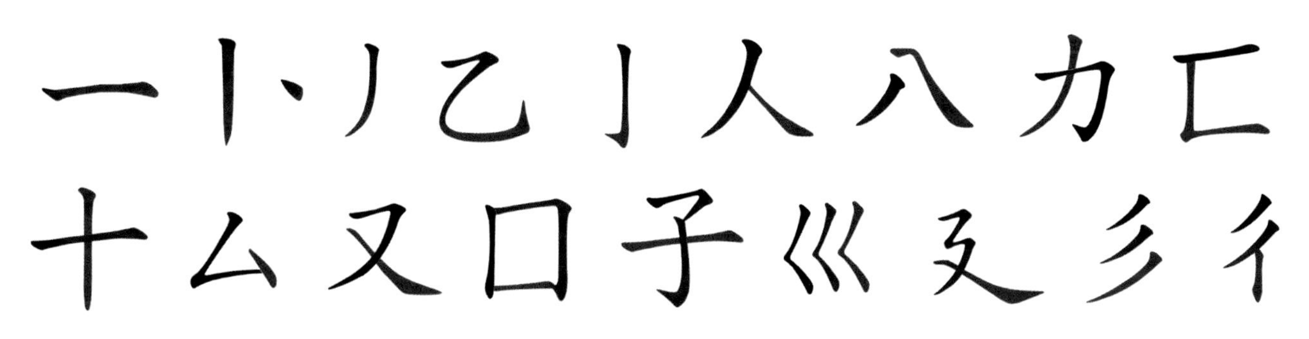

Flowing instructions at instantaneous point of need for brushstroke sequence flow width path.
Han characters depicted by **black = stroke-beginning** **red = stroke-ending**

Wikipedia Commons. Stroke Order Project

In Fort Mason on San Francisco Bay, people fish from piers with excellent views of Alcatraz. At this peaceful isolated place, San Mateo County Environmental Health Services provide instructions at point of need – warning about fish contaminated by PCBs and mercury.

MARKING UP THE WORLD AT IMMEDIATE POINT OF NEED

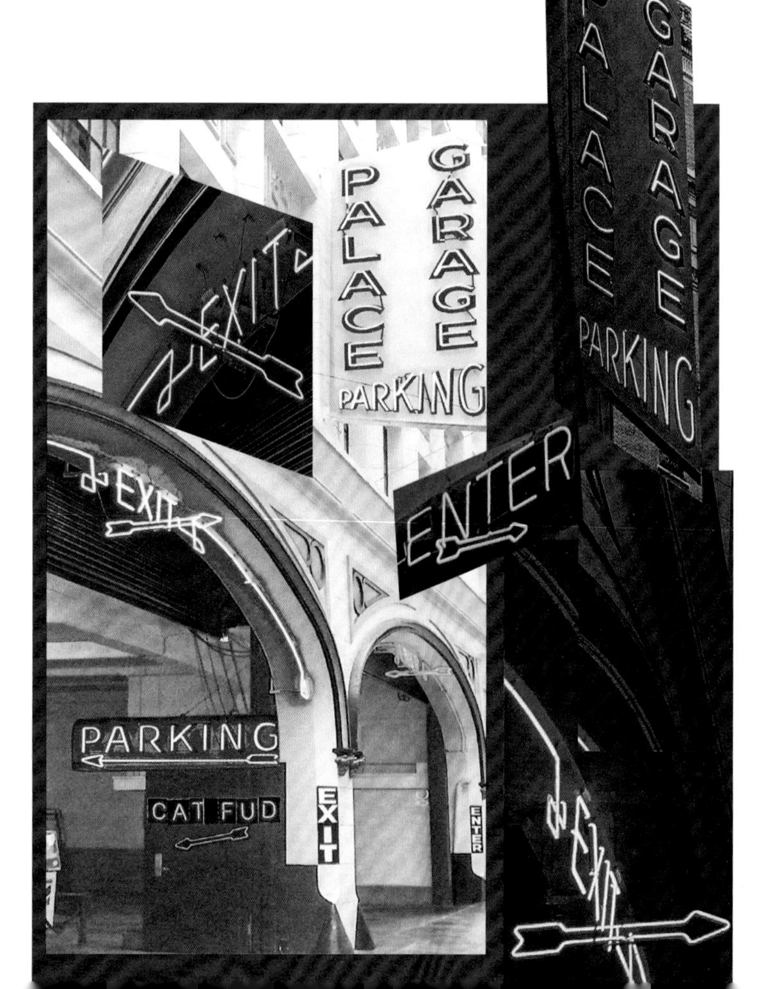

Collage of 3D pointers at the Palace Parking Garage, San Francisco

Interlinear text annotations at exact point of need: how to pronounce Roman numerals correctly when reading manuscript aloud ~1090 CE

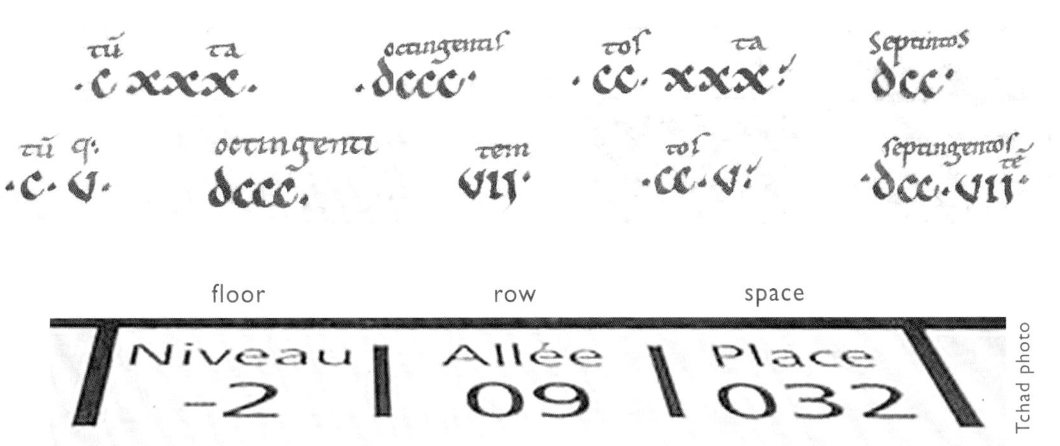

Parking-space address at Charles de Gaulle Airport, Paris
Car drivers take photographs to memorize where their car is parked

INSTRUMENT INVENTORY CONTROL DURING HEART SURGERY: CHECKLISTS AT POINT OF NEED

Real-time instrument inventory during heart surgery kept by single-patient-use regular-tip non-latex sterile surgical marker, in a stack list on surgical blanket, 4 x 15 cm

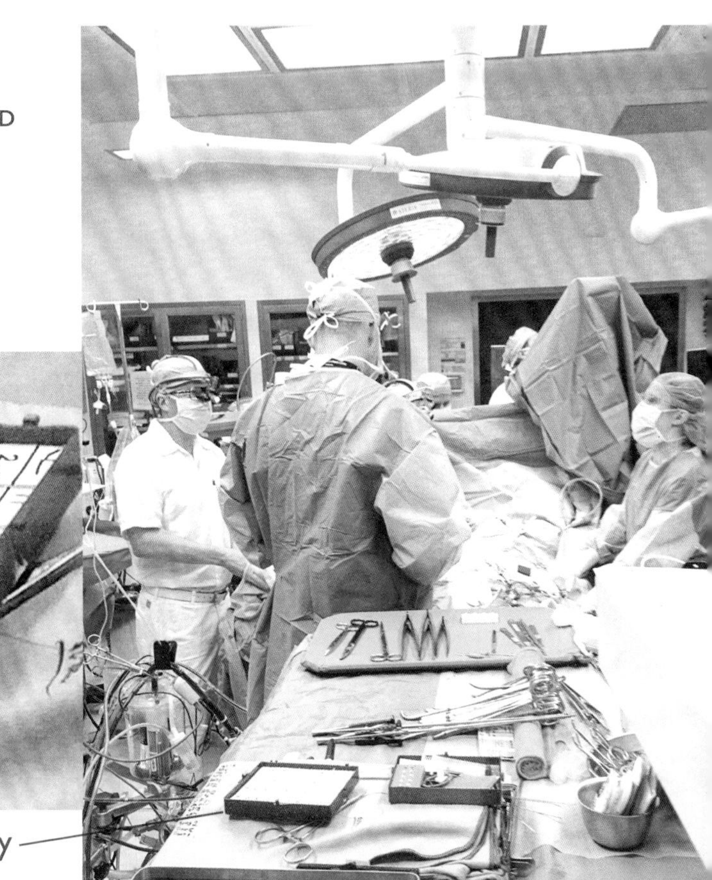

Foam-block needle count tray

INSTRUCTIONS AT POINT OF NEED FOR AUTHORS: PLOTCHARTS PLAN DETAIL, OVERVIEW, STRUCTURE

Preparing to write the novel *Catch 22*, Joseph Heller composed a storyboard, a 2-dimensional list with 3,650 words arrayed in 34 x 21 = 714 interacting cells. Rows are ordered in time, and each row records when each character does what. Some cell entries are erased. It took 7 years to complete the novel's 758-page typescript.

Laurence Sterne's *Tristram Shandy* (1767) depicts prankish plotcharts of wiggly lines, fully integrated into sentences (!), purporting to summarize the first 5 volumes – uproariously mocking over-wrought and clumsy signpost summaries:

> I Am now beginning to get fairly into my work; and by the help of a vegitable diet, with a few of the cold feeds, I make no doubt but I ſhall be able to go on with my uncle *Toby*'s ſtory, and my own, in a tolerable ſtraight line. Now,

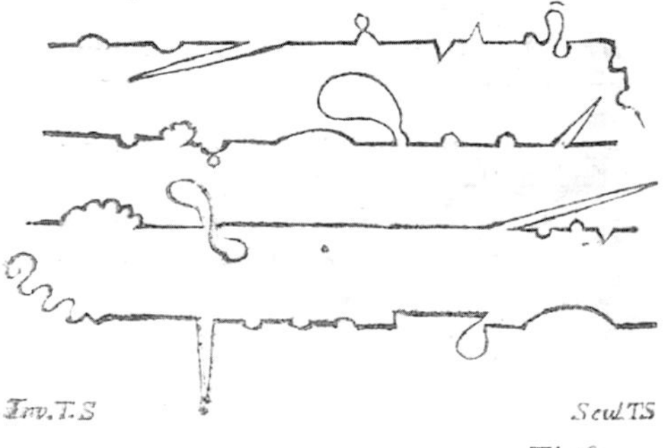

These

> Theſe were the four lines I moved in through my firſt, ſecond, third, and fourth volumes.——In the fifth volume I have been very good,——the preciſe line I have deſcribed in it being this :

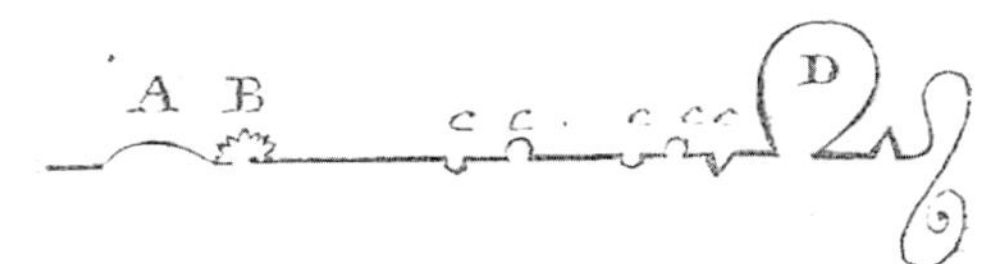

> By which it appears, that except at the curve, marked A. where I took a trip to *Navarre*,—and the indented curve B. which is the ſhort airing when I was there with the Lady *Bauſſiere* and her

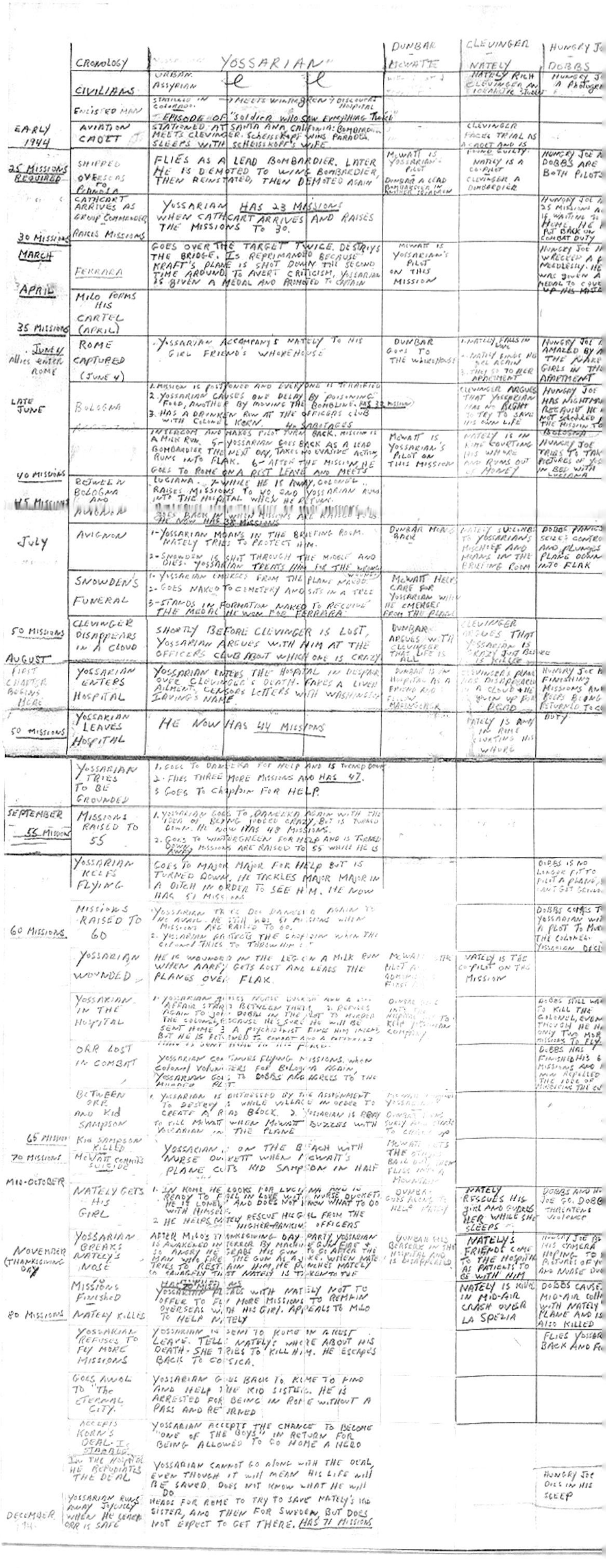

	CRONOLOGY	YOSSARIAN	DUNBAR – McWATTE	CLEVINGER – NATELY	HUNGRY JOE – DOBBS
	CIVILIANS	URBAN. ASSYRIAN	[illegible]	NATELY RICH. CLEVINGER AN IDEALISTIC STUDENT	HUNGRY JOE A PHOTOGRAPHER
	ENLISTED MAN	STATIONED IN COLORADO. → MEETS WINTERGREEN → DISCOVERS HOSPITAL. EPISODE OF "SOLDIER WHO SAW EVERYTHING TWICE"			
EARLY 1944	AVIATION CADET	STATIONED AT SANTA ANA CALIFORNIA: BOMBARDIER. MEETS CLEVINGER. SCHEISSKOPF WINS PARADES. SLEEPS WITH SCHEISSKOPF'S WIFE.		CLEVINGER FACES TRIAL AS A CADET AND IS FOUND GUILTY.	
25 MISSIONS REQUIRED	SHIPPED OVERSEAS TO PIANOSA	FLIES AS A LEAD BOMBARDIER. LATER HE IS DEMOTED TO WING BOMBARDIER, THEN REINSTATED, THEN DEMOTED AGAIN	McWATT IS YOSSARIAN'S PILOT. DUNBAR A LEAD BOMBARDIER IN [illegible]	NATELY IS A CO-PILOT. CLEVINGER A BOMBARDIER	HUNGRY JOE AND DOBBS ARE BOTH PILOTS
30 MISSIONS	CATHCART ARRIVES AS GROUP COMMANDER – RAISES MISSIONS	YOSSARIAN HAS 23 MISSIONS WHEN CATHCART ARRIVES AND RAISES THE MISSIONS TO 30.			HUNGRY JOE HAS 25 MISSIONS AND IS WAITING TO GO HOME. HE IS PUT BACK ON COMBAT DUTY
MARCH	FERRARA	GOES OVER THE TARGET TWICE. DESTROYS THE BRIDGE. IS REPRIMANDED BECAUSE KRAFT'S PLANE IS SHOT DOWN THE SECOND TIME AROUND. TO AVERT CRITICISM, YOSSARIAN IS GIVEN A MEDAL AND PROMOTED TO CAPTAIN	McWATT IS YOSSARIAN'S PILOT ON THIS MISSION		HUNGRY JOE HAS WRECKED A PLANE NEEDLESSLY. HE WAS GIVEN A MEDAL TO COVER UP HIS MISTAKE
APRIL / 35 MISSIONS	MILO FORMS HIS CARTEL (APRIL)				
JUNE 4 Allies enter ROME	ROME CAPTURED (JUNE 4)	YOSSARIAN ACCOMPANYS NATELY TO HIS GIRL FRIEND'S WHOREHOUSE	DUNBAR GOES TO THE WHOREHOUSE	1. NATELY FALLS IN LOVE. 2. NATELY FINDS HIS GIRL AGAIN. 3. THEY GO TO HER APARTMENT	HUNGRY JOE IS AMAZED BY THE NAKED GIRLS IN THE APARTMENT
LATE JUNE	BOLOGNA	1. MISSION IS POSTPONED AND EVERYONE IS TERRIFIED. 2. YOSSARIAN CAUSES ONE DELAY BY "POISONING" FOOD, ANOTHER BY MOVING THE BOMBLINE. HAS 32 MISSIONS. 3. HAS A DRUNKEN ROW AT THE OFFICERS CLUB WITH COLONEL KORN. 4. SABOTAGES		CLEVINGER ARGUES THAT YOSSARIAN HAS NO RIGHT TO TRY TO SAVE HIS OWN LIFE	HUNGRY JOE HAS NIGHTMARES BECAUSE HE IS NOT SCHEDULED FOR THE MISSION TO BOLOGNA
40 MISSIONS		INTERCOM AND MAKES PILOT TURN BACK. MISSION IS A MILK RUN. 5 – YOSSARIAN GOES BACK AS A LEAD BOMBARDIER THE NEXT DAY, TAKES NO EVASIVE ACTION, RUNS INTO FLAK. 6 – AFTER THE MISSION, HE GOES TO ROME ON A REST LEAVE AND MEETS	McWATT IS YOSSARIAN'S PILOT ON THIS MISSION	NATELY IS IN ROME COURTING HIS WHORE AND RUNS OUT OF MONEY	HUNGRY JOE TRIES TO TAKE PICTURES OF YOSSARIAN IN BED WITH LUCIANA
45 MISSIONS	BETWEEN BOLOGNA AND AVIGNON	LUCIANA. 7 – WHILE HE IS AWAY, COLONEL RAISES MISSIONS TO 40, AND YOSSARIAN RUNS INTO THE HOSPITAL WHEN HE RETURNS. [illegible] GOES BACK IN WHEN MISSIONS ARE RAISED TO 45. HE NOW HAS 38 MISSIONS			
JULY	AVIGNON	1 – YOSSARIAN MOANS IN THE BRIEFING ROOM. NATELY TRIES TO PROTECT HIM. 2 – SNOWDEN IS SHOT THROUGH THE MIDDLE AND DIES. YOSSARIAN TREATS HIM FOR THE WRONG WOUND	DUNBAR MOANS BACK	NATELY SUCCUMBS TO YOSSARIAN'S MISCHIEF AND MOANS IN THE BRIEFING ROOM	DOBBS PANICS, SEIZES CONTROL AND PLUNGES PLANE DOWN INTO FLAK
	SNOWDEN'S FUNERAL	1 – YOSSARIAN EMERGES FROM THE PLANE NAKED. 2 – GOES NAKED TO CEMETERY AND SITS IN A TREE. 3 – STANDS IN FORMATION NAKED TO RECEIVE THE MEDAL HE WON FOR FERRARA	McWATT HELPS CARE FOR YOSSARIAN WHEN HE EMERGES FROM THE PLANE		
50 MISSIONS / AUGUST	CLEVINGER DISAPPEARS IN A CLOUD	SHORTLY BEFORE CLEVINGER IS LOST, YOSSARIAN ARGUES WITH HIM AT THE OFFICERS CLUB ABOUT WHICH ONE IS CRAZY	DUNBAR ARGUES WITH CLEVINGER THAT LIFE IS ALL	CLEVINGER ARGUES THAT YOSSARIAN IS CRAZY JUST BEFORE [HE] IS KILLED	
FIRST CHAPTER BEGINS HERE	YOSSARIAN ENTERS HOSPITAL	YOSSARIAN ENTERS THE HOSPITAL IN DESPAIR OVER CLEVINGER'S DEATH. FAKES A LIVER AILMENT, CENSORS LETTERS WITH WASHINGTON IRVING'S NAME	DUNBAR IS IN HOSPITAL AS A FRIEND AND FELLOW MALINGERER	CLEVINGER'S PLANE HAS DISAPPEARED IN A CLOUD + HE [IS] GIVEN UP FOR DEAD	HUNGRY JOE FINISHING MISSIONS AND KEEPS BEING RETURNED TO COMBAT DUTY
50 MISSIONS	YOSSARIAN LEAVES HOSPITAL	HE NOW HAS 44 MISSIONS		NATELY IS AWAY IN ROME COURTING HIS WHORE	
	YOSSARIAN TRIES TO BE GROUNDED	1. GOES TO DANEEKA FOR HELP AND IS TURNED DOWN. 2. FLIES THREE MORE MISSIONS AND HAS 47. 3. GOES TO CHAPLAIN FOR HELP.			
SEPTEMBER – 55 MISSIONS	MISSIONS RAISED TO 55	1. YOSSARIAN GOES TO DANEEKA AGAIN WITH THE IDEA OF BEING JUDGED CRAZY, BUT IS TURNED DOWN. HE NOW HAS 48 MISSIONS. 2. GOES TO WINTERGREEN FOR HELP AND IS TURNED DOWN. MISSIONS ARE RAISED TO 55 WHILE HE IS AWAY			
	YOSSARIAN KEEPS FLYING	GOES TO MAJOR MAJOR FOR HELP BUT IS TURNED DOWN. HE TACKLES MAJOR MAJOR IN A DITCH IN ORDER TO SEE HIM. HE NOW HAS 51 MISSIONS			DOBBS IS NO LONGER FIT TO PILOT A PLANE, [BUT] CAN'T GET GROUNDED
60 MISSIONS	MISSIONS RAISED TO 60	1. YOSSARIAN TRIES DOC DANEEKA AGAIN TO NO AVAIL. HE STILL HAS 51 MISSIONS WHEN MISSIONS ARE RAISED TO 60. 2. YOSSARIAN [illegible] THE CHAPLAIN WHEN THE COLONEL TRIES TO THROW HIM OUT			DOBBS COMES TO YOSSARIAN WITH A PLOT TO MURDER THE COLONEL. YOSSARIAN DECLINES
	YOSSARIAN WOUNDED	HE IS WOUNDED IN THE LEG ON A MILK RUN WHEN AARFY GETS LOST AND LEADS THE PLANES OVER FLAK.	McWATT [IS THE] PILOT [illegible]	NATELY IS THE CO-PILOT ON THE MISSION	
	YOSSARIAN IN THE HOSPITAL	1. YOSSARIAN [illegible] NURSE DUCKETT AND A [illegible] AFFAIR STARTS BETWEEN THEM. 2. REFUSES AGAIN TO JOIN DOBBS IN THE PLOT TO MURDER THE COLONEL, BECAUSE HE'S SURE HE WILL BE SENT HOME. 3. A PSYCHIATRIST FINDS HIM INSANE BUT HE IS RETURNED TO COMBAT AND A [illegible] [illegible] IS SENT HOME IN HIS PLACE.	DUNBAR [illegible] INTO THE HOSPITAL [illegible] TO KEEP YOSSARIAN COMPANY		DOBBS STILL WANTS TO KILL THE COLONEL, EVEN THOUGH HE HAS ONLY TWO MORE MISSIONS TO FLY
	ORR LOST IN COMBAT	YOSSARIAN CONTINUES FLYING MISSIONS. WHEN COLONEL VOLUNTEERS FOR BOLOGNA AGAIN, YOSSARIAN GOES TO DOBBS AND AGREES TO THE MURDER PLOT.			DOBBS HAS FINISHED HIS 60 MISSIONS AND [illegible] REPELLED [BY] THE IDEA OF MURDERING THE COLONEL
	BETWEEN ORR AND KID SAMPSON	1. YOSSARIAN IS DISTRESSED BY THE ASSIGNMENT TO DESTROY A WHOLE VILLAGE IN ORDER TO CREATE A ROAD BLOCK. 2. YOSSARIAN IS READY TO KILL McWATT WHEN McWATT BUZZES WITH YOSSARIAN IN THE PLANE	McWATT [illegible] YOSSARIAN. DUNBAR [illegible] SURLY AND STARTS TO CRACK UP		
65 MISSIONS / 70 MISSIONS	KID SAMPSON KILLED. McWATT COMMITS SUICIDE	YOSSARIAN IS ON THE BEACH WITH NURSE DUCKETT WHEN McWATT'S PLANE CUTS KID SAMPSON IN HALF	McWATT LETS THE OTHERS BAIL OUT, THEN FLIES INTO A MOUNTAIN		
MID-OCTOBER	NATELY GETS HIS GIRL	1. IN ROME HE LOOKS FOR LUCIANA AND IS READY TO FALL IN LOVE WITH NURSE DUCKETT. HE IS LONELY AND DOES NOT KNOW WHAT TO DO WITH HIMSELF. 2. HE HELPS NATELY RESCUE HIS GIRL FROM THE HIGHER-RANKING OFFICERS	DUNBAR GOES ALONG TO HELP NATELY	NATELY RESCUES HIS GIRL AND GUARDS HER WHILE SHE SLEEPS	DOBBS AND HUNGRY JOE GO. DOBBS THREATENS VIOLENCE
NOVEMBER (THANKSGIVING DAY)	YOSSARIAN BREAKS NATELY'S NOSE	AFTER MILO'S THANKSGIVING DAY PARTY, YOSSARIAN IS AWAKENED IN TERROR BY MACHINE GUN FIRE + SO ANGRY HE GRABS HIS GUN TO GO AFTER THE MAN WHO FIRED THE GUN AS A JOKE. WHEN NATELY TRIES TO RESTRAIN HIM, HE PUNCHES NATELY SO SAVAGELY THAT NATELY IS TAKEN TO THE HOSPITAL	DUNBAR GOES BERSERK IN THE HOSPITAL AND IS DISAPPEARED.	NATELY'S FRIENDS COME TO THE HOSPITAL AS PATIENTS TO BE WITH HIM	HUNGRY JOE [BRINGS] HIS CAMERA HOPING TO [TAKE] PICTURES OF YOSSARIAN AND NURSE DUCKETT
80 MISSIONS	70 MISSIONS FINISHED. NATELY KILLED	HAS 70 MISSIONS. YOSSARIAN PLEADS WITH NATELY NOT TO OFFER TO FLY MORE MISSIONS TO REMAIN OVERSEAS WITH HIS GIRL. APPEALS TO MILO TO HELP NATELY		NATELY IS KILLED IN MID-AIR CRASH OVER LA SPEZIA	DOBBS CAUSES MID-AIR COLLISION WITH NATELY'S PLANE AND IS ALSO KILLED
	YOSSARIAN REFUSES TO FLY MORE MISSIONS	YOSSARIAN IS SENT TO ROME ON A REST LEAVE. TELLS NATELY'S WHORE ABOUT HIS DEATH. SHE TRIES TO KILL HIM. HE ESCAPES BACK TO CORSICA.			FLIES YOSSARIAN BACK AND [illegible]
	GOES AWOL TO "THE ETERNAL CITY."	YOSSARIAN GOES BACK TO ROME TO FIND AND HELP THE KID SISTER. HE IS ARRESTED FOR BEING IN ROME WITHOUT A PASS AND RETURNED			
	ACCEPTS KORN'S DEAL. IS STABBED.	YOSSARIAN ACCEPTS THE CHANCE TO BECOME "ONE OF THE BOYS" IN RETURN FOR BEING ALLOWED TO GO HOME A HERO			
	IN THE HOSPITAL HE REPUDIATES THE DEAL	YOSSARIAN CANNOT GO ALONG WITH THE DEAL, EVEN THOUGH IT WILL MEAN HIS LIFE WILL BE SAVED. DOES NOT KNOW WHAT HE WILL DO			HUNGRY JOE DIES IN HIS SLEEP
DECEMBER 1944	YOSSARIAN RUNS AWAY JOYOUSLY WHEN HE LEARNS ORR IS SAFE	HEADS FOR ROME TO TRY TO SAVE NATELY'S KID SISTER, AND THEN FOR SWEDEN, BUT DOES NOT EXPECT TO GET THERE. HAS 71 MISSIONS			

The *Catch 22* plotchart works better upon replacing optically noisy grids with ghost grids. Lightness of framing lines creates soft boundaries to maintain order and also allows words to spill across cells naturally. This small section at right mutes the dark grid clutter.

More generally, ask of information displays and interfaces, "What is the strongest visual element?" The correct answer is not "grid lines".

MILO	AARFY / CAPTAIN BLACK	CHAPLAIN	DOC DANEEKA / DR. STUBBS	OTHER COMBAT OFFICERS	ENLISTED MEN	MAJOR MAJOR / MAJOR DANBY	COLONEL CATHCART / COLONEL KORN	GENERAL DREEDLE / GENERAL PECKEM	ITALIANS	NURSES + OTHER AM. WOMEN	CASUALTIES	NOTES	THE WAR IN EUROPE
MARRIED; A FAITHFUL HUSBAND	AARFY A COLLEGE ALUMNUS	MARRIED: WITH YOUNG CHILDREN	DANEEKA: PRIVATE PRACTICE BEFORE WAR, KICKBACKS, ABORTIONS			MAJOR MAJOR LOOKS LIKE HENRY FONDA	CATHCART A HARVARD GRADUATE WITH A CIGARETTE HOLDER						1943 6/11 - SICILY LANDINGS
					WINTERGREEN ADVISES YOSSARIAN TO DO HIS DUTY	PROMOTED BY AN IBM MACHINE					SOLDIER WHO SEES EVERYTHING TWICE DIES OF SOME MYSTERIOUS DISEASE	FAMILY OF THE DEAD SOLDIER VISITS YOSSARIAN INSTEAD OF THEIR SON	9/9 - AMERICAN FORCES LAND AT SALERNO
MILO IS A PILOT						TRAINED AT CADET SCHOOL AS A PILOT				SCHEISSKOPF'S WIFE HAS AFFAIR WITH YOSSARIAN	[illegible]	SCHEISSKOPF HAS A MANIA FOR PARADES	
VOLUNTEERS FOR MESS OFFICER AS SOON AS HE ARRIVES OVERSEAS	AARFY IS YOSSARIAN'S NAVIGATOR. BLACK IS THE SQUADRON INTELLIGENCE OFFICER	THE CHAPLAIN DOES NOT FIT IN AND HAS NO FRIENDS					BECOMES A FULL COLONEL AT AGE 36. LONGS TO BE A GENERAL	DREEDLE HAS TAKEN HIS SON-IN-LAW, COLONEL MOODUS, INTO THE BUSINESS			COL. NEVERS: PREVIOUS GROUP COMMANDER IS KILLED OVER AREZZO.	CATHCART'S PREDECESSOR WAS, PRESUMABLY, A SATISFACTORY OFFICER WHO DID NOT SHIRK HIS COMBAT DUTIES.	1/22 - ANZIO LANDINGS
HE PRODUCES FRESH EGGS AND BEGINS ORGANIZING HIS BUYING SYNDICATE		HE FEARS BOTH COLONELS AND IS BULLIED BY THEM.				FINDS TRUE HAPPINESS AS A COMBAT PILOT, WHERE HE FINALLY FITS IN.	ARRIVES AS THE NEW GROUP COMMANDER AND RAISES MISSIONS TO 30					CATHCART ARRIVES AS THE NEW COMMANDER AND IMMEDIATELY RAISES THE MISSIONS TO 30	1944 JAN-MARCH: CASSINO BATTLES
	AARFY MISLEADS YOSSARIAN, MAKING IT NECESSARY FOR THEM TO GO AROUND TWICE			KRAFT IS KILLED ON YOSSARIAN'S SECOND BOMB RUN.			REPRIMANDS YOSSARIAN, THEN DECIDES TO PROMOTE HIM AND GIVE HIM A MEDAL				KRAFT: HIS PLANE BLOWS UP RIGHT AFTER YOSSARIAN'S SECOND BOMB RUN	CATHCART HAS VOLUNTEERED HIS MEN FOR FERRARA TO MAKE A GOOD IMPRESSION FOR HIMSELF. YOSSARIAN FEELS RESPONSIBLE FOR KRAFT'S DEATH	5/12 - ANZIO + CASSINO BREAKTHROUGH
MILO FLIES TO EGYPT AND BUYS THE WHOLE COTTON CROP	BLACK STARTS HIS LOYALTY OATH CRUSADE AGAINST MAJOR MAJOR	HE LOVES HIS WIFE AND MISSES HER.				MAJOR MAJOR IS APPOINTED SQD. COMMANDER WHEN PREDECESSOR IS KILLED OVER PERUGIA	DISAPPROVE OF BLACK'S LOYALTY OATH CAMPAIGN BUT AFRAID TO INTERVENE				MAJOR DULUTH: PREVIOUS SQUADRON COMMANDER IS KILLED IN ACTION OVER PERUGIA	MAJOR MAJOR'S PREDECESSOR, LIKE COLONEL CATHCART'S, IS KILLED IN COMBAT.	
MILO ARRANGES THE MISSION TO ORVIETO AS A BUSINESS DEAL, WORKING FOR BOTH SIDES	AARFY SPEAKS OF NATELY'S GIRL WITH SMUG CONTEMPT			MUDD, THE DEAD MAN IN YOSSARIAN'S TENT IS KILLED OVER ORVIETO THE DAY HE ARRIVES		AS SQD. COMMANDER HE IS OSTRACIZED AND ABUSED AND MADE THE VICTIM OF BLACK'S LOYALTY OATH CRUSADE			NATELY'S WHORE SITS NAKED IN A ROOM FULL OF ENLISTED MEN AND IGNORES HER		MUDD: THE DEAD MAN IN YOSSARIAN'S TENT IS KILLED OVER ORVIETO	MILO ORGANIZES THE MISSION TO ORVIETO AS A SOUND BUSINESS VENTURE AND DOES NOT FEEL GUILTY OVER MUDD'S DEATH.	6/4 - ROME ENTERED; 6/6 - NORMANDY LANDINGS
MILO HAS DISAPPROVED OF THE LOYALTY OATH CRUSADE BECAUSE IT DISRUPTED BUSINESS	BLACK REALIZES THAT MEN ARE GOING TO BOLOGNA AND HANDS THE [illegible] SIGN ON THE CLOSED MEDICAL TENT		STUBBS IS SORRY FOR THE MEN AND FEELS YOSSARIAN MIGHT BE THE ONLY SANE PERSON LEFT		WINTERGREEN REFUSES TO HELP AND LECTURES YOSSARIAN ON HIS DUTY TO BE KILLED	AFTER HE IS BEATEN UP ON THE BASKETBALL COURT HE TURNS HIMSELF INTO A RECLUSE AND SEES NO ONE	KORN CLOSES THE MEDICAL TENTS WHILE THE MEN ARE WAITING TO FLY TO BOLOGNA	PECKEM RECEIVES A MEDAL AND BEGINS RECOMMENDING THAT BOMBING ACTIVITIES BE PLACED UNDER HIS OWN SPECIAL SERVICE COMMAND	THE OLD MAN IN THE WHORE HOUSE BAITS NATELY. GIRL GETS BORED WITH ARGUMENT AND ABANDONS NATELY				
	AARFY GETS EVERYTHING WRONG ON THE BOLOGNA MISSION + DOESN'T HEAR WHEN YOSSARIAN SHOUTS AT HIM TO GET OUT OF THE NOSE				WINTERGREEN'S OWN DUTY, AS HE SEES IT, IS TO MAKE A PROFIT ON HIS BLACK MARKET OPERATIONS			PECKEM IS A SMOOTH POLITICIAN. DREEDLE IS A BLUNT OUTSPOKEN MAN	YOSSARIAN MEETS LUCIANA, THROWS AWAY HER ADDRESS. SLEEPS WITH MANY IN LIME-SCENTED [illegible] WHEN HE CANNOT FIND HER AGAIN		ANONYMOUS PLANES ARE SHOT DOWN ON THE MISSION TO BOLOGNA	YOSSARIAN THROWS AWAY LUCIANA'S ADDRESS AND SPENDS THE REST OF THE BOOK LOOKING AND LONGING FOR HER	
MILO BOMBS HIS OWN SQUADRON WHEN PAID TO DO SO BY THE GERMANS	BLACK SLEEPS WITH NATELY'S GIRL JUST TO TAUNT HIM	HE WORRIES CONSTANTLY ABOUT THE HEALTH AND SAFETY OF HIS WIFE AND CHILDREN	DANEEKA BEHAVES WITH COURAGE WHEN MILO BOMBS THE SQUADRON					WHO DRINKS HEAVILY AND KNOWS HE IS LOSING OUT BECAUSE HE DOES NOT MAKE THE RIGHT CONTACTS				MILO JUSTIFIES BOMBING THE SQUADRON IN TERMS OF FREE ENTERPRISE AND THE LARGE PROFIT HE HAS MADE	
MILO HAS REMOVED THE MORPHINE FROM THE FIRST AID KIT, AND THERE IS NOTHING TO EASE SNOWDEN'S PAIN			DANEEKA TAKES CARE OF YOSSARIAN WHEN HE EMERGES FROM THE PLANE IN A STATE OF SHOCK		SNOWDEN IS SHOT THROUGH THE MIDDLE AND DIES	DANBY FAINTS WHILE CONDUCTING BRIEFING AND IS ORDERED TAKEN OUTSIDE AND SHOT	KORN GRANDSTANDS FOR GENERAL DREEDLE BUT EARNS HIS CONTEMPT	GENERAL DREEDLE BRINGS HIS GIRL FRIEND TO THE AVIGNON BRIEFING	NATELY'S WHORE LIKES CAPTAIN BLACK AND GIVES HIM SOME OF THE MONEY NATELY PAYS HER		SNOWDEN "FREEZES TO DEATH" WHILE YOSSARIAN ATTENDS HIM		
MILO SEEKS YOSSARIAN OUT DURING SNOWDEN'S FUNERAL FOR HELP IN DISPOSING OF COTTON		CHAPLAIN SPIES YOSSARIAN NAKED IN A TREE AND THINKS IT IS A MYSTICAL VISION				BOTH MAJORS ATTEND SNOWDEN'S FUNERAL	CATHCART VOWS TO PUNISH YOSSARIAN FOR STANDING NAKED IN FORMATION	DREEDLE AWARDS MEDALS AND SIDES WITH YOSSARIAN WHEN CATHCART OFFERS TO PUNISH HIM					
											CLEVINGER: PLANE FLIES INTO CLOUD AND IS NEVER SEEN AGAIN		8/12 - FLORENCE CAPTURED; 8/15 - INVASION OF SOUTHERN FRANCE
MILO TRUSTS YOSSARIAN BECAUSE HE WILL NOT STEAL FROM THE COUNTRY HE LOVES	AARFY IS YOSSARIAN'S INCOMPETENT LEAD NAVIGATOR	THE CHAPLAIN MEETS YOSSARIAN IN THE HOSPITAL	DANEEKA IS A SELF-PITYING HYPOCHONDRIAC WITH WHOM YOSSARIAN IS FRIENDLY	THE SOLDIER IN WHITE APPEARS IN THE HOSPITAL WARD AND IS JUDGED DEAD	SNOWDEN HAS ALREADY BEEN KILLED. WINTERGREEN, AS MAIL CLERK, IS ALREADY INFLUENTIAL	MAJOR MAJOR IS VISITED BY CID MAN AND BEGINS TO SIGN PAPERS WASHINGTON IRVING	CATHCART HAS RAISED MISSIONS FROM 25 TO 50	DREEDLE AND PECKEM ARE FEUDING AND VYING FOR POWER	NATELY'S WHORE IS BORED WITH HIM AND PAYS HIM LITTLE ATTENTION	YOSSARIAN HAS NO RELATIONSHIP WITH NURSE DUCKETT YET	SOLDIER IN WHITE PRONOUNCED DEAD IN THE HOSPITAL	FIRST CHAPTER BEGINS HERE	8/23 - PARIS LIBERATED; 8/23 - RUSSIA OCCUPIES RUMANIA
MILO IS AWAY BUYING FIGS							CATHCART STRIVES TO IMPRESS BOTH GENERALS FAVORABLY	NEITHER GENERAL TAKES ENOUGH NOTICE OF CATHCART TO INTERVENE EITHER WAY				MEN HAVE TO FLY 50 MISSIONS, ALTHOUGH HEADQUARTERS REQUIRES ONLY 40.	
		CHAPLAIN HAS NO AUTHORITY TO HELP; BUT THEY BECOME FRIENDS	DANEEKA REFUSES TO HELP BUT SAYS HEADQUARTERS WANTS ONLY 40 MISSIONS			MAJOR MAJOR VISITED BY SECOND CID MAN AND PUTS HIM ON THE TRAIL OF THE FIRST							9/11 - ALLIED FORCES MOVE INTO GERMANY
MILO ADVISES HIM TO FLY THE 55 MISSIONS LIKE A GOOD SOLDIER			EXPLAINS CATCH-22 IN TERMS OF ORR		WINTERGREEN REFUSES TO HELP AND WILL TURN AGAINST HIM IF HE GETS INTO TROUBLE							~~FIRST CHAPTER BEGINS HERE~~	9/21 - ALLIES PUSH PAST PISA
			DANEEKA SHARES YOSSARIAN'S FEAR OF DEATH AND WILL TAKE NO CHANCES OF ANY KIND			MAJOR MAJOR IS SORRY FOR YOSSARIAN BUT THERE IS NOTHING HE CAN DO	BOTH COLONELS ARE INVOLVED IN "PLUM TOMATO" DEALS WITH MILO						
	BLACK TAUNTS NATELY AT THE OFFICERS CLUB ABOUT SCREWING HIS GIRL	SUMMONED BY COLONEL TO SAY PRAYERS IN THE BRIEFING ROOM. COMPLAINS IN VAIN ABOUT THE 60 MISSIONS	DR. STUBBS DOES GROUND THE MEN IN HIS SQUADRON, EVEN THOUGH KORN PUTS THEM RIGHT BACK ON COMBAT DUTY				COLONEL WANTS PRAYERS IN THE BRIEFING ROOM IN ORDER TO GET INTO SATURDAY EVENING POST						10/2 - SIEGFRIED LINE COLLAPSES, RUSSIA OCCUPIES BULGARIA
	AARFY IS UNABLE TO HEAR YOSSARIAN'S SCREAMS OR SEE HIS BLOOD	TRIES TO SEE MAJOR MAJOR ON YOSSARIAN'S BEHALF	DANEEKA AGAIN REFUSES TO GROUND YOSSARIAN, EVEN THOUGH HE HAS BEEN JUDGED CRAZY				COLONEL IS TERRIFIED BY THE MENTION OF YOSSARIAN'S NAME						
MILO HAS REMOVED THE GAS CYLINDERS FROM THE MAE WESTS TO MAKE ICE CREAM SODAS FOR THE OFFICERS		CHAPLAIN SEES DANEEKA ON YOSSARIAN'S BEHALF + IS GOING TO TRY WINTERGREEN NEXT			WINTERGREEN SENDS THE CHAPLAIN AWAY WITHOUT SEEING HIM					YOSSARIAN GOOSES NURSE DUCKETT			11/15 - AMERICANS APPROACH BOLOGNA
				SCHEISSKOPF IS SHIPPED OVERSEAS TO GENERAL PECKEM			CATHCART VOLUNTEERS FOR BOLOGNA AGAIN	PECKEM SETS UP CAMPAIGN AGAINST DREEDLE		THEY BECOME LOVERS.	ORR DITCHES AND IS GIVEN UP FOR LOST		
	BLACK'S TAUNTS CAUSE CHAPLAIN TO DOUBT HIS WIFE'S FIDELITY						COLONELS ARE NOT CONCERNED WITH THE ROAD BLOCK. THEY WANT A GOOD AERIAL PHOTOGRAPH			SHE TRIES TO REFORM HIM			
			DANEEKA IS OFFICIALLY "DEAD", WHEN MCWATT'S PLANE CRASHES				USES THESE DEATHS AS AN EXCUSE TO RAISE MISSIONS TO 70			SHE IS ON THE BEACH WITH HIM WHEN KID SAMPSON IS KILLED	KID SAMPSON CUT IN HALF. MCWATT FLIES INTO MOUNTAIN	DANEEKA, BY REFUSING TO GET INVOLVED IN ANYTHING, HAS BEEN "DEAD" ALL THE TIME	
	AARFY DOES NOT GET ALONG BECAUSE HE DOES NOT WANT TO GET IN BAD WITH ANYONE		DR. STUBBS IS DEFYING GROUP + KORN WANTS TO GET RID OF ALL FLIGHT SURGEONS						NATELY'S WHORE GETS A GOOD NIGHT'S SLEEP AND FALLS IN LOVE WITH HIM			AS SOON AS NATELY GETS HIS GIRL HE TURNS INSUFFERABLY BOURGEOISE AND TRIES TO REFORM EVERYBODY	
MILO GIVES A HUGE BANQUET AT WHICH EVERYONE GIVES THANKS TO HIM		EVEN THE CHAPLAIN ENTERS THE WARD PLEASED HE HAS AT LAST HAD THE COURAGE TO SIN			SGT. KNIGHT FIRES A MACHINE GUN AT NIGHT AND SHOUTS "HAPPY NEW YEAR!"				NATELY'S WHORE IS FURIOUS WITH YOSSARIAN FOR HITTING HIM	WARNS YOSSARIAN THAT THEY ARE GOING TO DISAPPEAR DUNBAR	DUNBAR: HE IS DISAPPEARED	NURSE DUCKETT HAS BEGUN TO THINK OF HER FUTURE. SHE WANTS TO MARRY A DOCTOR AND HAS BEGUN TO DISASSOCIATE HERSELF FROM YOSSARIAN, WHOM SHE DOES NOT REGARD AS A SUITABLE HUSBAND	12/16-1/20, 1945 BATTLE OF THE BULGE
MILO PERSUADES CATHCART TO RAISE MISSIONS TO 80, KNOWING NATELY WILL FLY THEM		THE CHAPLAIN IS TAKEN TO A CELLAR, TRIED, AND FOUND GUILTY. ANGRILY, HE DECIDES TO GO DIRECTLY TO GENERAL DREEDLE	DR. STUBBS IS TRANSFERRED TO THE PACIFIC	SCHEISSKOPF FINDS HIMSELF IN CHARGE OF EVERYTHING	WINTERGREEN BLAMES PECKEM WHEN SCHEISSKOPF IS PUT IN CHARGE OF THEM ALL			PECKEM REPLACES DREEDLE. THEN HIS OWN RECOMMENDATION IS ADOPTED + HE FINDS HIMSELF UNDER SCHEISSKOPF			NATELY, DOBBS: KILLED IN MID-AIR CRASH	MILO AND CATHCART HAVE EACH FLOWN NO MORE THAN FIVE MISSIONS	3/7 - REMAGEN BRIDGE CAPTURED; 4/1 - RUSSIANS ENTER AUSTRIA
	BLACK LAUGHS WHEN WHOREHOUSE IS RAIDED AND GIRLS ARE CHASED OUT			THE OTHER MEN SIDE WITH YOSSARIAN SECRETLY BUT NONE WILL JOIN HIM				PECKEM SENDS THE CHAPLAIN AWAY WITHOUT SEEING HIM	NATELY'S WHORE TRIES TO STAB YOSSARIAN AND BEGINS HER CHASE AFTER HIM	NURSE DUCKETT IS EMBARRASSED BY YOSSARIAN'S STAND		NATELY'S WHORE BECOMES A SYMBOL OF HIS GUILT AND RESPONSIBILITY FOR NEVER INTERVENING IN THE INJUSTICES HE KNOWS EXIST EVERYWHERE	5/3 - BERLIN FALLS; GERMANS SURRENDER IN ITALY
MILO URGES HIM PIOUSLY TO DO HIS DUTY. DESERTS HIM TO SMUGGLE TOBACCO	AARFY RAPES AND MURDERS THE MAID BUT IS ALLOWED TO GO FREE								KID SISTER IS ADRIFT IN ROME WITH NO ONE TO HELP HER		OLD MAN: DIES OF OLD AGE. MAID: MURDERED BY AARFY		
		CHAPLAIN DOES NOT KNOW WHAT YOSSARIAN SHOULD DO. CHAPLAIN IS DELIGHTED AND ENCOURAGED BY YOSSARIAN'S DECISION.		THE OTHER MEN WOULD BE TRICKED BY YOSSARIAN'S DECISION AND INSPIRED TO CONTINUE FLYING	YOSSARIAN HAS TREATED SNOWDEN FOR THE "WRONG" WOUND.	DANBY COMES TO PERSUADE YOSSARIAN TO GO THROUGH WITH THE DEAL BUT IS CONVERTED TO YOSSARIAN'S POINT OF VIEW. DANBY AIDS HIM IN ESCAPE AND OFFERS MONEY AND ADVICE	COLONELS MUST DEAL WITH YOSSARIAN OR PUNISH HIM		NATELY'S WHORE STABS YOSSARIAN AS HE LEAVES THE OFFICE. NATELY'S WHORE IS RIGHT BEHIND YOSSARIAN WITH A KNIFE	NURSE DUCKETT HELPS HIM ESCAPE BECAUSE SHE IS EAGER TO BE RID OF HIM	HUNGRY JOE: DIES IN HIS SLEEP	IN MAKING THE DECISION TO DESERT, YOSSARIAN ACCEPTS THE RESPONSIBILITY HE NOW KNOWS HE HAS TO THE OTHER MEN. AS HE SAYS, HE IS NOT RUNNING AWAY FROM HIS RESPONSIBILITIES, BUT TO THEM	

FRESH EYES: SEE NOW . . . WORDS LATER

INSTRUCTIONS FOR SEEING WITH FRESH EYES

Word-guided seeing blocks fresh seeing. Suppose this artwork appeared with a museum description, *A Rare Byzantine Orthodoxium Sacred Cross* (a strong narrative for many). Or the title is *Hommage in Steel for JC* (a reference to Joseph Cornell's collage boxes filled with bird images)? Or the news is out that the artwork sold for 1.2 million euros (now it looks precious indeed)? Or a curator denounced it as a 19th-century fake (oops)? Or it was made from the bars of torture cells melted down by freed political prisoners? Or that the piece celebrates or ironicizes an open-ended implement wrench, an *objet trouvé*? Or how about *The Chicken Goes up the Hill,* a title making it impossible to see anything other than an inclined chicken climb?

Abstract sculptors make objects generating *unique optical experiences*. These one-off experiences exist utterly independent of artchat. Abstract artworkers often insist "our only language is vision."

Seen real time by viewers walking around outdoors, abstract works provide a multiplicity of direct and vivid optical experiences: form, scale, color, shadows, volume, airspace, landscape. Direct optical experiences are universal, produced by nature's universal laws that determine how light bounces around 3D artworks. Light from abstract sculptures is focused by the lens of the eye onto the retina. Optic nerves link retinal images to the brain and download optical information at a combined 500,000 bits per second. What then?

These two pages are revised from ET, *Seeing Around* (2010)

Our minds are quick to convert new optical experiences into familiar stories, favored viewpoints, comforting metaphors. No wonder, for how else can we manage optical data flows at 500,000 bits per second without a million predetermined categories for filing, without the rage for wanting to conclude?

Pre-installed narratives, categories, stale metaphors, reminiscences, *déformation professionnelle* all interfere with how and what we see. It's confirmation bias, all over again. In looking at abstract artworks, once words and story-telling start, it's hard to see anything else. Old words deform new seeing. Hamlet's words dominate the seeing of the suggestible Polonius:

Hamlet	*Do you see yonder cloud that's almost in shape of a camel?*
Polonius	*By the mass. And 'tis like a camel indeed.*
Hamlet	*Methinks it is like a weasel.*
Polonius	*It is backed like a weasel.*
Hamlet	*Or like a whale?*
Polonius	*Very like a whale.*

To see with fresh, uninstructed eyes and an open mind requires a deliberate, self-aware act by the observer. Abstract artworks represent themselves and should be first viewed for themselves. When looking at outdoor abstract pieces, concentrate initially on the *unique optical experience* produced by the artworks. See as the artist saw when making the piece.

A focus on optical experience does not deny stories, it postpones them. Viewing an artwork may evoke interesting narratives – or just tedious artchat recalling similar art or artists, concocting playful tales, realizing how scrap metal was repurposed into art, making judgements about the artist's intentions or character, or contemplating an artwork's provenance, price, politics. Let the artwork stand on its own. Walk around fast and slow, be still, look and see from up down sideways close afar above below, enjoy the multiplicity of silhouettes shadows dapples clouds airspaces sun earth glowing. Your only language is vision.

The rage for wanting to conclude is one of the most deadly and most fruitless manias to befall humanity. Each religion and each philosophy has pretended to have God to itself, to measure the infinite, and to know the recipe for happiness. What arrogance and what nonsense! I see, to the contrary, that the greatest geniuses and the greatest works have never concluded.

Gustave Flaubert, *Correspondance* (Paris, 1929), V, 111.

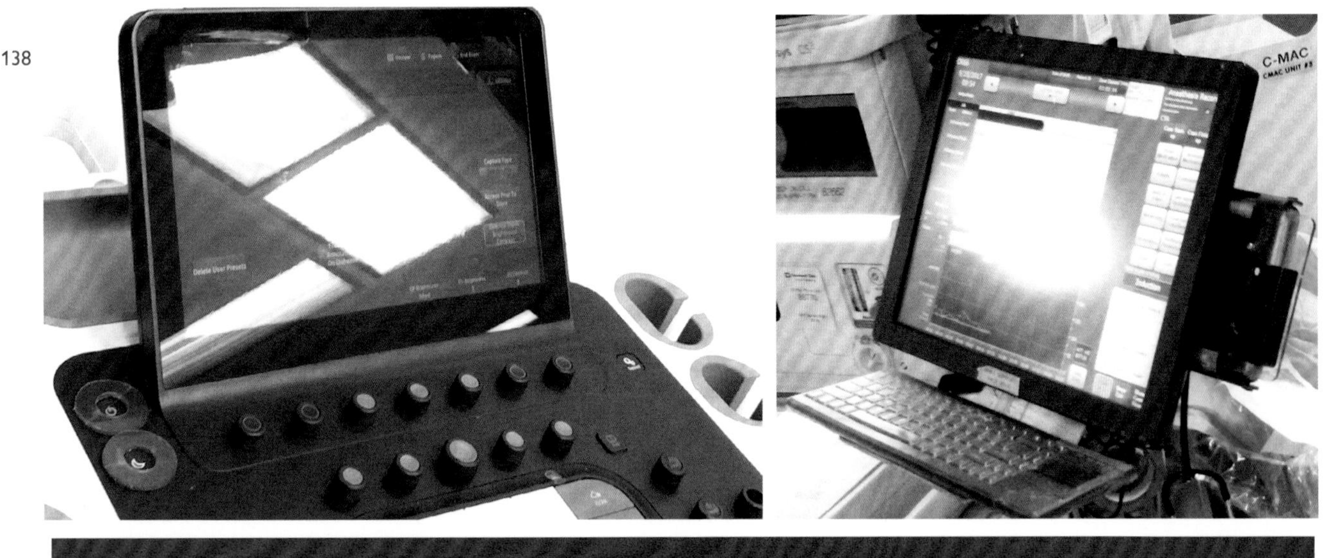

utilities > setups > system settings > display > control panel brightness

OPTICAL NOISE COMPROMISES VISUAL SIGNALS. REFLECTIONS AND GLARE EXIST WHERE LIGHT EXISTS

CAN VIEWERS ADAPT/ACCOMMODATE? CAN GLARE BE REDUCED AT POINT OF NEED?

Ceiling-light reflections show up on glossy screens for control panels in heart surgery operating rooms. Reflections can be blown out by increasing screen brightness but not here: brightness controls are buried 5 levels deep. Chasing optical noise by increasing brightness may convert screens into *sources* of glare – as in noisy restaurants where people talk louder because people are talking louder. Everyone taking photographs with a camera phone in sunlight must manage fierce reflections from glossy display screens.

When do display screens in serious places deserve NASA-level detail, care, knowledge? Does it matter? Compared to flight decks, glare and reflections in operating rooms are harder to calm down. New sources of optical noise show up every time a new piece of surgical equipment arrives. Here are instructions by Asaf Degani for NASA flight-deck information on paper. These instructions are relevant for the *much greater glare from glossy laptop and iPad screens now used for documents and flight maps.*

"**4.1 GLARE** Reference handbooks and checklists used by airlines and military units are laminated to protect them from wear and tear. Others are inserted into a plastic casing and are pulled out only when a new revision is issued. In choosing a plastic cover or lamination, *an anti-glare plastic that diffuses light is recommended; otherwise, some rays from the light source will be reflected to the pilot's eyes.* This is commonly observed in a dim cockpit when the pilot's light is directed to a document covered with glossy plastic. Other types of glare occur during night operation. When printed matter is lit by direct light and the pilot's eyes shift between the document and darkened windows or control panels, the eyes must constantly re-adapt to different levels of luminance. Severe differences in luminance between the document in front of the observer (critical vision) and the surroundings (peripheral vision), reduce visual discrimination, reading speed, comfort. Furthermore, any strong light source (direct sunlight, radar scope) that is not shielded from the field of vision, will cause disability glare. As light sources get closer to the line-of-sight an increased reduction in visual efficiency will be experienced."

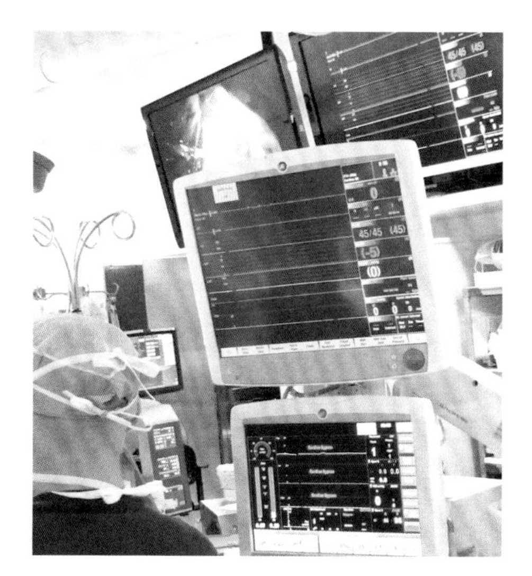

Working with multiple screens requires aligning multiple viewing angles to suppress multiple reflections. Users can accommodate themselves, but attention to screen design and placement can reduce such optical artifacts. This high-level operating room in 2017 employed 25 display screens, 22 glossy and 3 matte. In exchange for glare reduction, matte screens slightly fog images. For important viewing, such as pre-op echocardiograms, display screens can be placed in alcoves providing optical and acoustical serenity. Flat data-lines on these screens indicate the heart and lungs are stopped, and the patient is on the heart-lung machine.

Movement attracts the attention of human eye-brain systems. As viewers move, glare and reflections move; 2 video frames show moving reflections from storage cabinets. This occurs everywhere there is glass – and everyone adapts. (Glass covered artworks, however, are noisome, at least for the hypervigilant.)

Glass and stainless steel storage cabinets generate 8 m^2 of optical noise as if reflected from dull distorting mirrors. Polished stainless steel on the cabinets accidentally creates anisotropic light; such polishing is intentionally used in stainless steel artworks to enliven beautiful light reflected off the artwork.

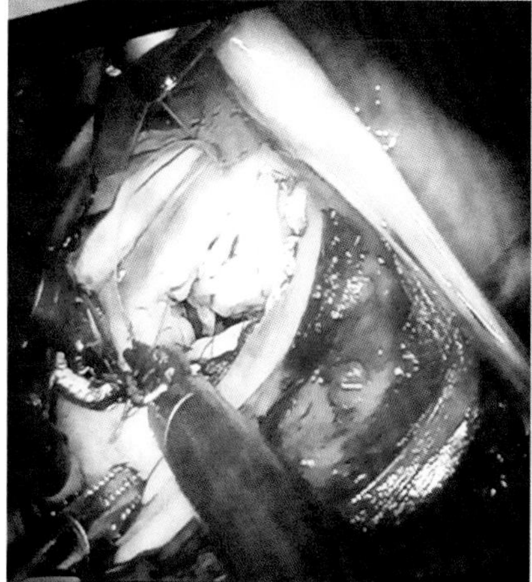

Here is a still image from the surgeon's live video view of a mitral valve repair inside the heart during robotic surgery: lights, scissors, scalpels, suction, needles are controlled by the surgeon and staff. Reflections come from wet surfaces and surgical instruments; the white specks are reflections from fat in the heart. In robotic surgery, the surgeon views the video feed in a darkened booth on a display screen free of reflections – unlike conventional glossy display screens that reflect surrounding light, such as ceiling lights in operating rooms.

This report is based on observing 3 heart surgeries (2 robotic, 1 open heart) done by the amazing Dr. Marc Gillinov at the amazing Cleveland Clinic. Three years earlier, my mitral valve was repaired in the same operating room. ET

Braque, baroque, barrack,
bark, poodle, Suzanne R., 68th St.? REgent 7-12..?, BUtterfield 8,
ALgonquin 4, ELdorado 5, El Morocco, Mogador, Mogadiscio, Abys-
-sinia, 1936!, Vittorio Emanuele III° Re d'Italia e di Albania Imperatore
d'Etiopia, George V, Louis XIV, Louis XVII, Louis XXXIX, Paris XIVe,
N.Y. 21, 22, 28, 17, 5, Monte Carlo, Monte Cristo, Montealegre, Bernstein,
Lev Davidovich Bronstein, Trotsky, ПРАВДА, Iskra, Herzen, Lunacharsky,
Stamboliski, Bakunin, Kropotkin, Kostrowitzki, Kandinsky, Kafka, Kupka,
Kokoschka, K, K, K, K, K,
AIR MAIL
RF
BLACK, BLUE, YELLOW,
RED, GREEN, Greenberg, Monteverdi, Verdi, Rossini, Leoncavallo, Catfish,
Ratfink, Schweinehunde, Dragonfly, Horsefly, Belmont, Jamaica,
Auteuil, San Siro, My Old Man, The Killers, Kilimanjaro
Kilogram, Kilometer, 5/8 Mile, Isotta-Fraschini, Hispano-
Suiza, Svizzera, Trieste, Joyce, James Joyce, Greta Garbo,
Donald Duck, BB, MM, Phileas Fogg, Ugene Unesco,
Tristan Tzara, Tară, Țața, Uța, Ața, Ița, None, Papa,
Gigi, Tata, Dada, Ada, Hedda, Betty Parsons,
Curt Vallentin, Maeght, Janis, Museum, Rockefeller,
Nelson, David, Hare, Denise, The Knees, Haircut,
Nosedrop, Gogol, Nabokov, Hi Nabor, While-U-Wait,
U Turn, U Thant, H. B.4. B.U.T., No X-ing, Vietato Fumare,
Defense d'Afficher, English Spocken, Inghiliss, Pidgin,
Bêche-de-Mer, Peles Kanaka, Kaikai liklik dok,
Em I nabisman tasol, Mi save tokboy, Talkboy, Thinkboy
STEINBERG

Saul Steinberg, cover, *The New Yorker*, October 18, 1969.

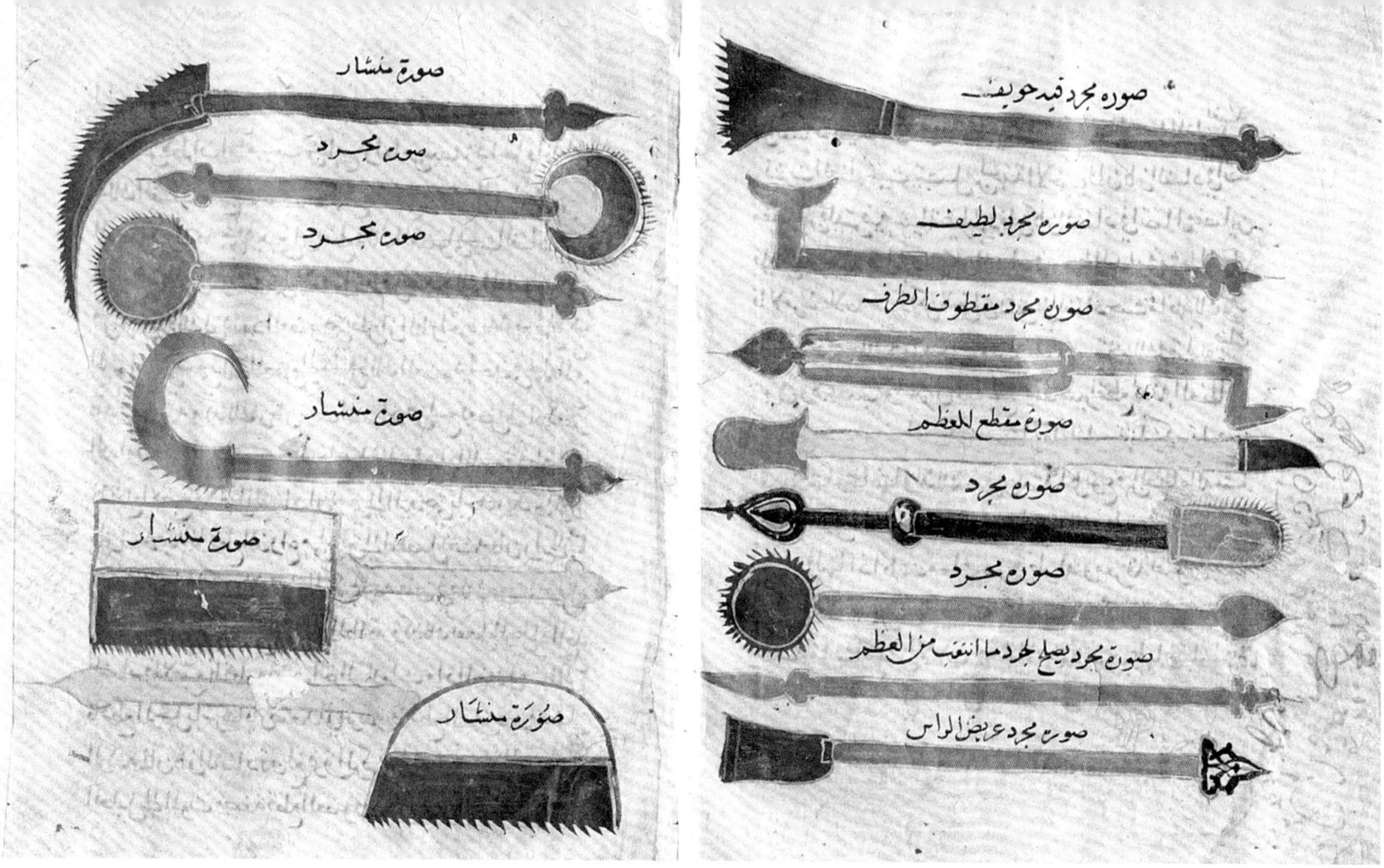

7 LISTS: THEORY AND PRACTICE

Lists consist of whatever it takes – nouns, proper nouns, verbs, graphics, images, numbers. Saul Steinberg's art list, free associations upon seeing a painting by Braque, jumbles words, typography, colors, sketches. Above, an Arabic medical textbook from ~1000 CE shows 2 visual-verbal lists integrating illustrations and words to describe surgical instruments.

Thousands of years of list-making have produced sorted elements, rigid arrays, catalogs, matrices, strings, complex communications among lists, multi-dimensional lists beyond linear strings and stacks, web searches returning lists of useful and gamed links, finite and infinite lists. And personal lists: informal, intense, routine, to-do, self-help, resolutions.

Despite this great mishmash of lists, it is possible to specify an empirical theory of lists. In lists, spaces have meaning, locating elements in relation to other elements. Lists are often free and independent from conventional rules of style-sheets/grammar/typography/punctuation. Lists also help us escape from the personal internalized mash-up style sheets of every writer and reader – a continuous low-level background buzz checking to see if word usage, spelling, punctuation, grammar are 'correct'. Lists are content – about the substance contained, not the container. An empirical theory for reasoning about lists includes

freedom from conventional typographic consciousness
selection of list items   list quality and completeness
models and conventions   comparing list architectures
order and sorting of list elements   2-dimensional arrays

Lists are based on policies of inclusion/exclusion. Parodying list membership choices, Jorge Luis Borges cooked up a fanciful Chinese encyclopedia *The Heavenly Emporium of Benevolent Knowledge* that sorted animals into 14 eccentric categories:

those that belong to the emperor
embalmed ones
those that are trained
suckling pigs
mermaids
fabulous ones
stray dogs
those included in this classification
those that tremble as if they were mad
innumerable ones
those drawn with a very fine camel hair brush
et cetera
those that have just broken a flower vase
those that look like flies from a long way off

Borges mocks the category problem in list-making: *lists are constructed by cherry-picking and lemon-dropping elements from ill-defined populations.* List elements suffer from double selection bias: the bias in selecting elements for the list at hand, and the bias resulting from a population of elements that survived long enough to become candidates for the new list. When lists matter, as in checklists for pilots and surgeons, list-makers and list-users should contemplate the consequences, mysteries, and foolings around of element inclusion/exclusion.

That lists may fail to achieve perfect completeness rarely matters, except possibly to logicians and post-modernists. Even inexact lists are useful. Would airborne logicians prefer that pilots not use flight checklists, because such lists are paradoxical or logically imperfect? Practical and wise list-makers say 'We seek clarity in list inclusion and exclusion principles. In practice, our checklists have measurably improved over previous methods. Good enough is often good enough.' Besides, the Goddess of Perfection is busy with more important things.

On constructing a safety checklist at the start of surgery, see Atul Gawande, *The Checklist Manifesto: How to Get Things Right.* Delightful collections of lists and commentary: Shaun Usher, *Lists of Note,* Francis Spufford, *The Chatto Book of Cabbages and Kings: Lists in Literature,* and Umberto Eco, *The Infinity of Lists.*

HIERARCHICAL LISTS: AGGRESSIVE, IMPOSED, COMPLICATED, OFTEN AN INCOMPREHENSIBLE MAZE TO OUTSIDERS. TESTING HIERARCHICAL MODELS IS EASY BUT RARELY DONE: MAKE DIRECT COMPARISONS WITH NONHIERARCHICAL MODELS OF INFORMATION

Martin Hardee Tufte Story: AnswerBook

A long, long time ago, my friend Eric Bergman and I were working on the user interface for *Sun AnswerBook,* which was a CD-based predecessor to *docs.sun.com*. Sun had invited Edward Tufte in to teach a session in our Boston office about the Grand Vistas of Information Architecture or something. After the class, we lured him into our usability lab to look at the user interface for Answerbook, of which we were very proud.

That was waaay before the web or HTML or PDF; AnswerBook was cool stuff: It consisted of a hierarchical topic chooser and query engine... that served as a remote control to a PostScript-based browser... that displayed whatever section of whatever document you chose in the chooser... all documents were written in troff or Frame or Interleaf and then published into PostScript via ditroff or fmbatch... and we hacked PostScript, inserted metadata and linked comments... and there was an object link resolver underneath that mapped object IDs to the appropriate PostScript pages/sections. Of course we had NeWS so PostScript rendering happened for free.

Anyway... we were very proud of our user interface and the fact that we had a way to browse 16,000 (!!) pages of documentation on a CD-ROM. But browsing the hierarchy felt a little complicated to us.

So we asked Tufte to come in and have a look, and were hoping perhaps for a pat on the head or some free advice. He played with our AnswerBook for 90 seconds, turned around, pronounced his review:

> "Dr. Spock's *Baby and Child Care* is a best-selling owner's manual for the most complicated 'product' imaginable – and it has only 2 levels of headings. You have 8 levels of hierarchy and I haven't stopped counting yet. No wonder you think it's complicated."

Oh.

LIST MODELS ASSESSED BY DIRECT COMPARISON

AnswerBook and *Baby and Child Care* both array complicated content, handled in very different ways. Hierarchical content piles up complexity, difficult for outsiders and even insiders to understand.

LIST INDICATORS IN SENTENCES

An inline list, 7 lines long, spaces out 6 list elements by 3 dots + 1 space, for example,

ditroff or fmbatch... and we hacked

In this 9th-century Latin manuscript, long spaces, not commas, cushion and announce list elements:

NATURE | MAN
WILD | TAME
COUNTRY | CITY

MIRACLE | MASTERY
CREATURE | CREATOR
CHANCE | CONTROL
ACCIDENT | PURPOSE

LIFE | ART
NATURALISM | SYMBOLISM
REALITY | [illegible]
REAL | IDEAL

DIONYSIAN | APOLLONIAN
VITALITY | MEASURE
INSTINCT | INTELLECT

SPACE | TIME
OBJECT | SUBJECT

EARTH | HEAVEN
FLESH | SPIRIT
EGO | SUPER-EGO

ROMANTIC | CLASSIC
MYSTICAL | LOGICAL
POETIC | SCIENTIFIC
IRREGULAR | MATHEMATICAL
DRAMA | ORDER

EXPRESSIONIST | CONSTRUCTIVIST
FAUVIST | CUBIST
SURREAL | ABSTRACT
INTRA-SUBJECTIVE | NON-OBJECTIVE
SPONTANEOUS | DELIBERATE
TAOIST | CONFUCIST

ACTUALITY | POSSIBILITY
NECESSITY | FREEDOM

VARIETY OF DUALISMS

In these intriguing two-dimensional lists on a postcard written by Ad Reinhardt, stacklisted words contemplate *dualisms*, more exquisite/subtle than *opposites*. Some 32 word pairs in 18 lists in 9 matrices combine to reveal both horizontal and vertical meanings/readings.

Images and drawings on this grid read across 4 rows (each stop-action pair), down 2 columns (sequence of actions).

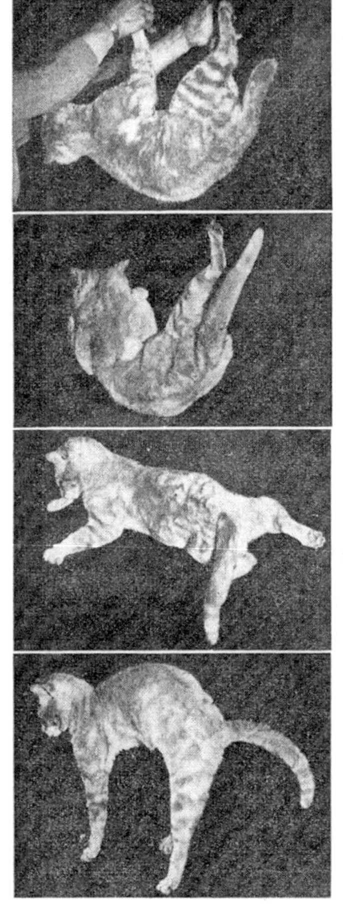

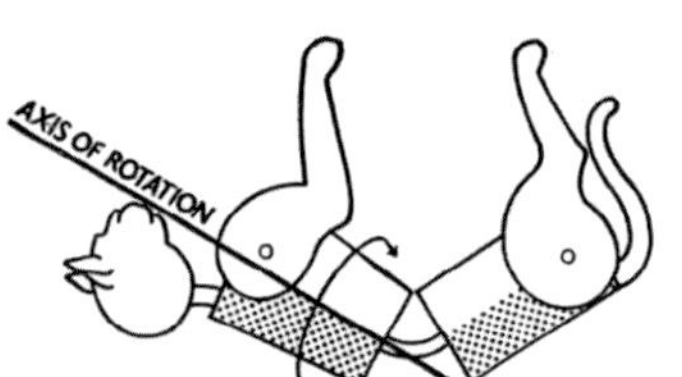

Flexion at waist. Fore-part begins to turn.

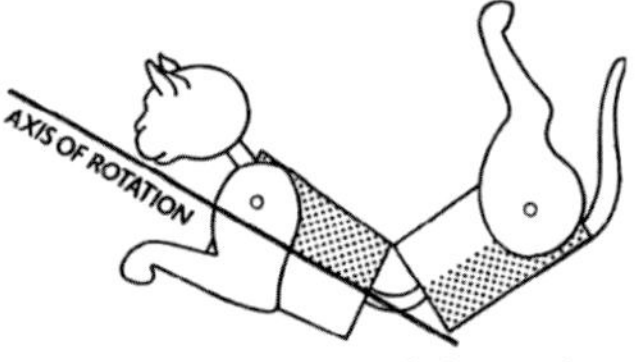

Fore-part rotated through 180°.

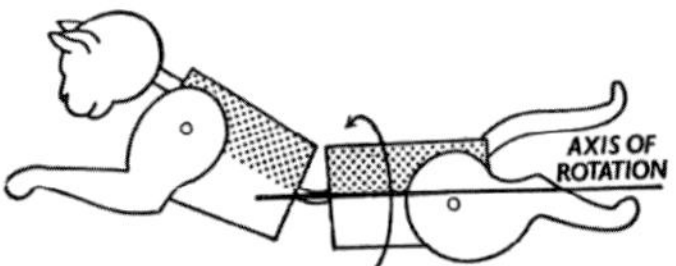

Rear end rotating on fore-end.

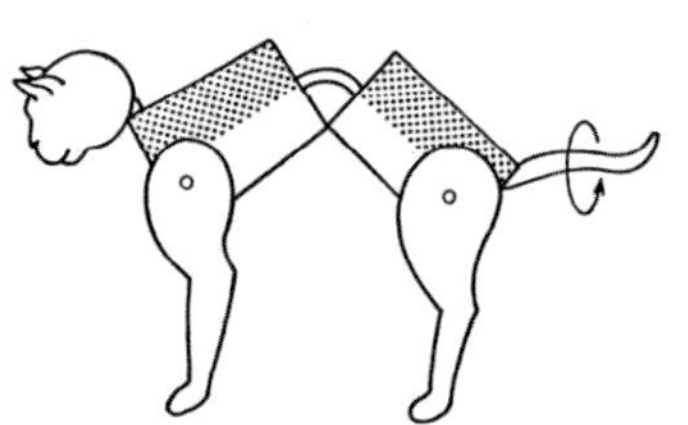

Back arched. Legs extended for landing. Tail circling to trim for leveling off.

A simple and natural design. Viewers see and understand how cats land on their feet.

TWO-DIMENSIONAL STACKLISTS CONNECTED AND INTERACTING

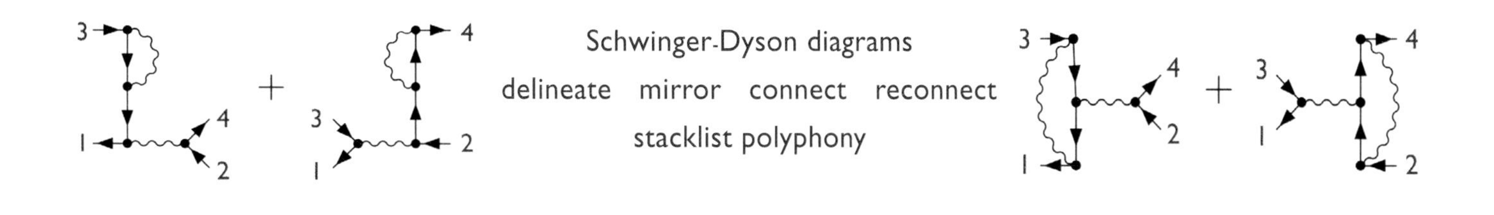

4 LISTS CONTAINING 19 SENTENCES LOCATED IN 2 BY 2 ARRAY

JACKIE HEINRICHER, *ON DISCOVERING BAMBOO*

CLUMPING BAMBOO

ADVANTAGES VIRTUES
Does not spread indefinitely throughout the soil.
Does not need physical barrier systems to control root growth.
Can provide evergreen screening throughout the year.
Plants range in height from 6 to well over 20 feet for temperate species (tropical species can range from 20 to well over 70 feet tall).
Base diameter ranges from 18 inches to 6 feet depending on species. Be aware that "open clumpers" can take up as much as 10 to 20 feet in diameter.
Requires water only at the base of the clump, not the entire planting area.

DISADVANTAGES VICES
Most temperate clumping bamboo have a cane diameter of no more than one-half inch.
Most temperate clumping bamboo will only reach 6 to 20 feet tall.
Many of the mountain clumpers will not tolerate high heat and humidity.

CLUMPING BAMBOO

RUNNING BAMBOO

ADVANTAGES VIRTUES
Wide range of plant selection.
Cane diameter can range from one-eighth inch to well over 6 inches.
Stabilizes soils and can be used for erosion control.
Provides an endless supply of craft materials.
New shoots can be edible depending on species.
Can be used as a wind break.
Sound effects included with each grove (excluding the groundcovers, which don't have much of a "rustling").

DISADVANTAGES VICES
Rhizomes can spread indefinitely in the soil.
Needs some kind of containment to control rhizome growth.
Can sneak in the windows at night and strangle the children.

RUNNING BAMBOO

(A joke, mocking nativist alarm of alien plants.)

A LIST OF 46 VERBS REMODEL IDEAS ABOUT CAUSALITY IN ART: AN ANNUNCIATION AND REVELATION

In an exquisitely divergent analysis, art historian Michael Baxandall suggests that newer artists act on and remodel the work of previous artists – reversing the usual model which claims that older influenced newer.

A bright-eyed list of 46 verbs, defining broad arrays of actions, flipping conventional models of agency, vigorously demonstrating that some models are better than others:

'Influence' is a curse of art criticism primarily because of its wrong-headed grammatical prejudice about who is the agent and who is the patient: it seems to reverse the active/passive relation which the historical actor experiences and the inferential beholder will wish to take into account.

If one says that *X influenced Y,*
it does seem that one is saying that
X did something to Y rather than that *Y did something to X.*
But in the consideration of good pictures and painters,
Y did something to X is always the more lively reality,
If we think of *Y rather than X* as the agent,
the vocabulary is much richer:

drawn on *resort to* *address* *assimilate* *simplify*
avail oneself of *copy* *paraphrase* *reconstitute*
appropriate from *absorb* *elaborate on*
have recourse to *make a variation on* *develop* *face up to*
adapt *misunderstand* *extract from* *master* *subvert*
refer to *revive* *continue* *perpetuate*
pick up *take on* remodel *ape* *reduce*
engage with *emulate* *travesty* *promote*
react to *quote* *parody* *distort* *respond to* *transform*
differentiate oneself from *attend to* *align oneself with*
assimilate oneself to *resist* *tackle*

RICHARD SERRA

TO-DO LIST

84 verbs

24 contexts

Verb List, 1967-1968

to roll
to crease
to fold
to store
to bend
to shorten
to twist
to dapple
to crumple
to shave
to tear
to chip
to split
to cut
to sever
to drop
to remove
to simplify
to differ
to disarrange
to open
to mix
to splash
to knot
to spill
to droop
to flow

to curve
to lift
to inlay
to impress
to fire
to flood
to smear
to rotate
to swirl
to support
to hook
to suspend
to spread
to hang
to collect
of tension
of gravity
of entropy
of nature
of grouping
of layering
of felting
to grasp
to tighten
to bundle
to heap
to gather

to scatter
to arrange
to repair
to discard
to pair
to distribute
to surfeit
to complement
to enclose
to surround
to encircle
to hole
to cover
to wrap
to dig
to tie
to bind
to weave
to join
to match
to laminate
to bond
to hinge
to mark
to expand
to dilute
to light

to modulate
to distill
of waves
of electromagnetic
of inertia
of ionization
of polarization
of refraction
of simultaneity
of tides
of reflection
of equilibrium
of symmetry
of friction
to stretch
to bounce
to erase
to spray
to systematize
to refer
to force
of mapping
of location
of context
of time
of carbonization
to continue

THE REMARKABLE STORY OF TWO DIVERGENT TEXT LISTS

Two lists populate this paragraph below from F. Scott Fitzgerald's novel *Tender is the Night* (1934). Then, at right on the adjacent page, David Lodge, a novelist and author of *The Art of Fiction* (1992), shows how meaning in Fitzgerald's list flows from words, syntax, structure, sequence.

F. Scott Fitzgerald

TENDER IS THE NIGHT

With Nicole's help Rosemary bought two dresses and two hats and four pairs of shoes with her money. Nicole bought from a great list that ran two pages, and bought the things in the windows besides. Everything she liked that she couldn't possibly use herself, she bought as a present for a friend. She bought colored beads, folding beach cushions, artificial flowers, honey, a guest bed, bags, scarfs, love birds, miniatures for a doll's house and three yards of some new cloth the color of prawns. She bought a dozen bathing suits, a rubber alligator, a travelling chess set of gold and ivory, big linen handkerchiefs for Abe, two chamois leather jackets of kingfisher blue and burning bush from Hermes—bought all these things not a bit like a high-class courtesan buying underwear and jewels, which were after all professional equipment and insurance—but with an entirely different point of view. Nicole was the product of much ingenuity and toil. For her sake trains began their run at Chicago and traversed the round belly of the continent to California; chicle factories fumed and link belts grew link by link in factories; men mixed toothpaste in vats and drew mouthwash out of copper hogsheads; girls canned tomatoes quickly in August or worked rudely at the Five-and-Tens on Christmas Eve; half-breed Indians toiled on Brazilian coffee plantations and dreamers were muscled out of patent rights in new tractors—these were some of the people who gave a tithe to Nicole, and as the whole system swayed and thundered onward it lent a feverish bloom to such processes of hers as wholesale buying, like the flush of a fireman's face holding his post before a spreading blaze. She illustrated very simple principles, containing in herself her own doom, but illustrated them so accurately that there was grace in the procedure, and presently Rosemary would try to imitate it.

TAKING LISTS SERIOUSLY: ANALYSIS OF THE TWO LISTS

DAVID LODGE
THE ART OF FICTION

"F. Scott Fitzgerald emphasizes the *miscellaneousness* of the list to convey the completely non-utilitarian nature of Nicole's shopping. Cheap, trivial things like coloured beads, and homely things like honey, are mixed up promiscuously with large functional objects like the bed, expensive toys like the gold-and-ivory chess set, and frivolities like the rubber alligator. There is no logical order in the list, no hierarchy of price, or importance, or grouping of the items according to any other principle. That is the point.

Nicole quickly exceeds the parameters of the list she brought with her, and buys whatever takes her fancy. By exercising her taste and gratifying her whims without regard to economy or commonsense, she conveys a sense of personality and temperament that is generous, impulsive, amusing, and aesthetically sensitive, if out of touch with reality in some important aspects. It's impossible not to respond to the fun and the sensual pleasure of this spending spree. How covetable those two chamois-leather jackets sound, kingfisher blue and burning bush (but the key word is 'two': where lesser mortals might hesitate between two identical jackets of different, equally attractive colours, Nicole solves the problem by buying both). No wonder her young protégé, and future rival, Rosemary, will try to imitate her style.

Balancing the shopping list, however, is another list, of the human beings, or groups, on whose exploitation Nicole's inherited wealth depends, a list which throws our response into reverse. The whole passage turns on the sentence, 'Nicole was the product of much ingenuity and toil,' which suddenly makes us see her not as the consumer and collector of commodities, but as herself a kind of commodity — the final, exquisite, disproportionately expensive and extravagantly wasteful product of industrial capitalism.

Whereas the first list was a sequence of *nouns,* the second is a series of *verb phrases:* 'trains began their run...chicle factories fumed...men mixed toothpaste...girls canned tomatoes...' At first sight these processes seem as mutually incongruous and randomly selected as the items of Nicole's shopping, but there is a connection between the men in the toothpaste factories and the girls in the dime stores and the Indian workers in Brazil: the profits made from their labour indirectly fund Nicole's shopping."

fire severity   alarm true   alarm false   inspection pass/fail   emergency

CASE STUDY: FIRE CHIEF MAKES PRESENTATION AT TOWN COUNCIL MEETING

At Town Council meetings, the Fire Department Chief reports monthly about fires, alarms, false alarms, inspections, expenditures. How to present 100s of data points and budget numbers with 10 other items on the agenda? Instead of a slide show, the Chief hands out a paper map that shows monthly activities, and budget line-item expenditures on the reverse side. This handout goes to Town Council members, news reporters, citizens at the meeting. The Fire Chief begins 'Look at our monthly report, then we will talk about it.' After reading, council members speak up: 'Why the Marina false alarms?' 'Why so few Southend inspections, so many fires?' 'What caused the Trois Marie fires?' 'What about these Lyon travel expenses?' Documents live on, unlike slides. Each Council member can look at any part of the report at any time, including the area that they represent and perhaps even their neighborhood. Maps make things personal for everyone.

CASE STUDY: HANDY-DANDY HANDOUTS FOR THE JURY

In a comic opera spat between Google and Uber over trade secrets and 'intellectual property,' Federal District Judge William Alsup, experienced in trials involving the angry rich, sought to speed up the case by providing the jury with paper handouts to keep in hand for reference during a complex case:

> "One of the things that I think the jury ought to have is a handy-dandy list, a list of alleged trade secrets," he said. The judge also urged the two sides to submit a glossary of the top dozen people so that jurors could have a clear notion of who is who.
>
> Cyrus Farivar, *ArsTechnica*, January 30, 2018

Presenters, your audience seeks to learn *What is the substantive content? What are the reasons to believe your presentation?* To improve presentations, improve the quality, relevance, integrity of your content. Prepare a credible document, a coherent series of reasons, facts, data. Present lots of signal, and no noise. If your numbers are boring, don't decorate them, get better numbers. Documents model information better than slide decks. To prove to yourself and others that you know what you are talking about, you should be able to write paragraphs describing

What the problem is.
Why the problem is relevant, why anyone should care.
How you're going to solve the problem.

Credibility derives from showing good evidence, comparing various viewpoints, briefly demonstrating mastery of detail, and avoiding biz-school/military/hi-tech jargon buzz words. Credibility is a continuing reputation for honest communication *and* getting it right. A good way to have credibility with your colleagues is not to have lied to them last week.

Presentations should now and then provide *quotations from experts* (mention their credentials). This helps the audience see technical language used correctly. Experts can express strong views, say things presenters can't possibly say, and demonstrate what intellectual leadership looks like. Use quotes *directly relevant* to the topic at hand, not faux-inspirational BS cheerleading.

A fundamental quality control mechanism for integrity is documentation: authors named, sponsors revealed, conflicts of interests and agendas unveiled, measurements verified. Deceptive documentation may disqualify presenters from their jobs. Every paragraph, every visualization should provide reasons to believe.

THINKING AND CARING ABOUT YOUR AUDIENCE

It's not what you say, it's what they hear. RED AUERBACH

The biggest problem in communication is the illusion that it has taken place. GEORGE BERNARD SHAW

Unnecessary noise is the most cruel absence of care inflicted on sick or well. FLORENCE NIGHTINGALE

Think the best you possibly can of your audience. Respect your audience – after all, they were good enough show up. Walk in the shoes of your audience. (Deflect, with civility, the occasional random jerks.) Do not spin, pander, dumb things down, use biz-school jargon. Your job is to get the content right, make everyone smarter, be honest, and finish early.

PRESENTERS: SHOW UP EARLY. FINISH EARLY.

Show up early to your own presentation. Early arrival may deflect minor problems, or you can chat with people in the audience. *Finish early,* 5% to 20% early. Leave them wanting more. The audience will be delighted. No one has ever complained about a presentation ending early. A good way to finish early is to stay off the long and winding road of slide decks.

PRACTICE YOUR PRESENTATION. EVEN BETTER, DON'T MAKE A PRESENTATION, INSTEAD PREPARE A DOCUMENT FOR A SILENT READING PERIOD FOLLOWED BY DISCUSSION.

A grand truth about human behavior is that *rehearsal improves performance.* During rehearsal, have a colleague from whom you can take criticism make comments. Identify and fix problems. Although nervous-making, rehearsals improve your presentations. A rehearsal or performance video will ruthlessly reveal incoherence, nervous habits. You can completely avoid slide decks, rehearsing, and presenting by beginning with a silent reading period of a document, followed by discussion! Prepare and rehearse for what might happen in the discussion.

TRUTH IS OFTEN COMPROMISED

Confirmation bias is the tendency to search for, interpret, favor, and recall information so as to confirm one's pre-existing beliefs or hypotheses. WIKIPEDIA *It is a principle that shines impartially on the just and the unjust alike that once you have a point of view all history will back you up.* VAN WYCK BROOKS

Although we often hear that data speak for themselves, their voices can be soft and sly. It is easy to lie with statistics; it is easier to lie without them. FREDERICK MOSTELLER

Bias can creep into the scientific enterprise in all sorts of ways. But financial conflicts are detectable definitively and represent a uniquely perverse influence on the search for scientific truth. COLIN BEGG

It is simply no longer possible to believe much of the clinical research that is published, or to rely on the judgment of trusted physicians or authoritative medical guidelines. I reached this conclusion slowly and reluctantly during my 2 decades as an Editor of The New England Journal of Medicine as drug companies asserted more power, and began to treat researchers as hired hands. MARCIA ANGELL

AUDIENCE RESPONSIBILITIES: TREAT PRESENTERS AS YOU WOULD LIKE TO BE TREATED

When you attend a presentation, stay on the content. See with an open mind but not an empty head. Be able to change your mind. Ignore slide design, decoration, chartjunk, or debating whether bullets should be gold stars ☆ or smiley faces 😊. Give your undivided attention to the content as long as you can, loot presentations for useful material, make the best of it. Learn something new, different, ingenious.

Don't let a few contrary elements in a presentation spoil your seeing and learning. If you require perfect agreement with presenters, you foreclose learning anything new, and you might as well stay home and stare at your immutable self in the mirror all day long. The purpose of a meeting is to solve problems. Listen, see, think, learn. Just because presenters disagree with the third paragraph of your budget statement doesn't mean they are Satanic. *The motives of your colleagues are usually no better or worse than your own.* This is true on average, and is also empirically true because people in the room are very much alike, due to prior selection, filtering, bias. With notable exceptions, your colleagues are *no more or less* intelligent, competent, open-minded, scheming, malicious, or Satanic than you are.

TO HAVE AN
OPEN MIND
BUT NOT AN
EMPTY HEAD
– TUFTE

AUSTIN KLEON

In talks, print out your notes/graphics/slides on a paper handout (A3 or 11 x 17 printed on both sides) adding up to 40 readable images. Here is my handout for a keynote on the future of data analysis and machine learning to 900 people. The paper serves as a nice take-away document, prompting memories of the talk. Some people in the audience will look it over before the talk starts, many will follow along. This document shows, for free, *other relevant material I thought the audience should learn about, but not presented in the talk itself.*

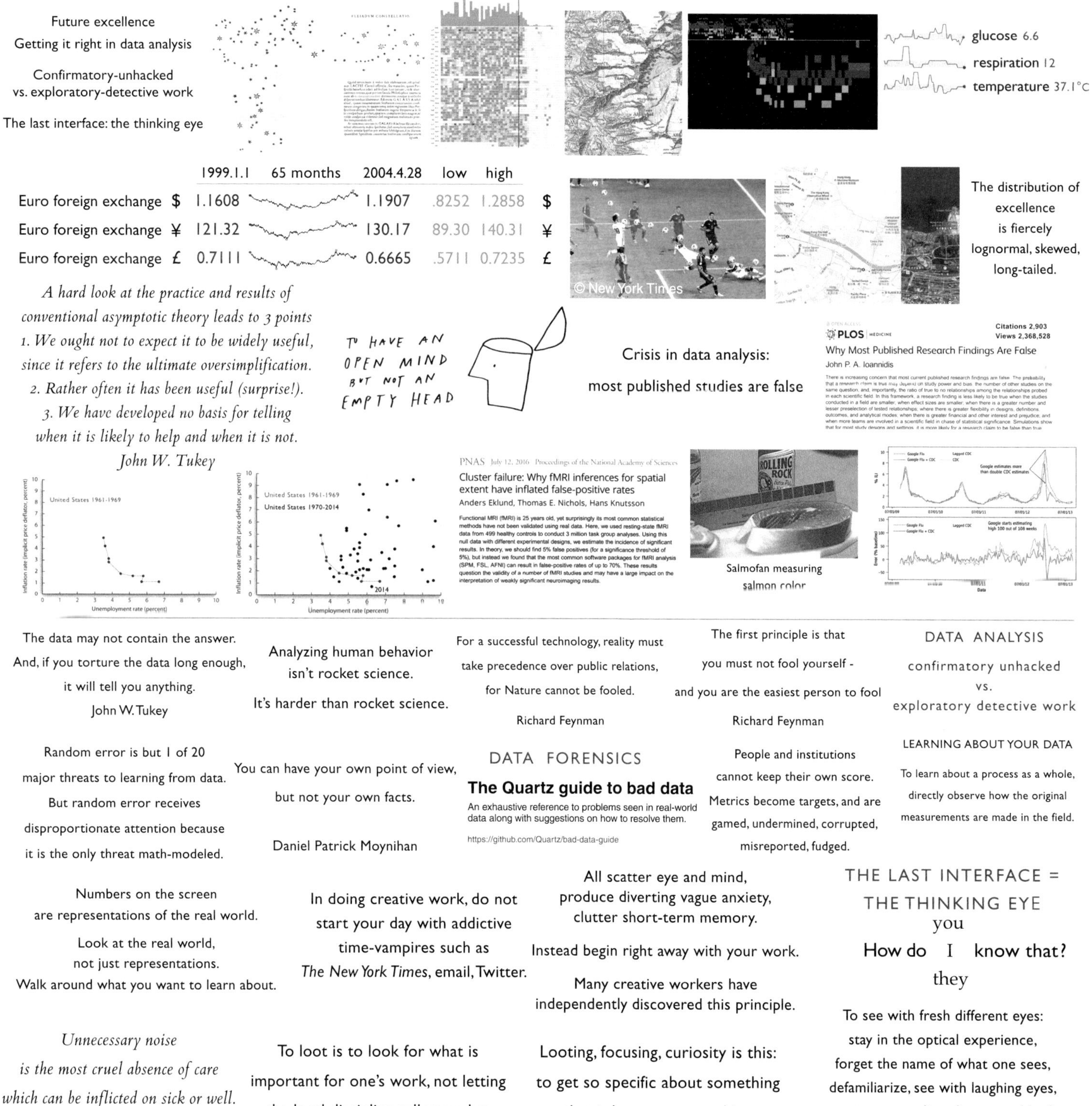

Future excellence
Getting it right in data analysis
Confirmatory-unhacked
vs. exploratory-detective work
The last interface: the thinking eye
glucose 6.6
respiration 12
temperature 37.1°C
1999.1.1 65 months 2004.4.28 low high
Euro foreign exchange $ 1.1608 1.1907 .8252 1.2858 $
Euro foreign exchange ¥ 121.32 130.17 89.30 140.31 ¥
Euro foreign exchange £ 0.7111 0.6665 .5711 0.7235 £
© New York Times
The distribution of excellence is fiercely lognormal, skewed, long-tailed.
A hard look at the practice and results of conventional asymptotic theory leads to 3 points
1. We ought not to expect it to be widely useful, since it refers to the ultimate oversimplification.
2. Rather often it has been useful (surprise!).
3. We have developed no basis for telling when it is likely to help and when it is not.
John W. Tukey
TO HAVE AN OPEN MIND BUT NOT AN EMPTY HEAD
Crisis in data analysis:
most published studies are false
PLOS MEDICINE
Citations 2,903
Views 2,368,528
Why Most Published Research Findings Are False
John P. A. Ioannidis
United States 1961-1969
Unemployment rate (percent)
Inflation rate (implicit price deflator, percent)
United States 1961-1969
United States 1970-2014
2014
PNAS
Cluster failure: Why fMRI inferences for spatial extent have inflated false-positive rates
Anders Eklund, Thomas E. Nichols, Hans Knutsson
Functional MRI (fMRI) is 25 years old, yet surprisingly its most common statistical methods have not been validated using real data. Here, we used resting-state fMRI data from 499 healthy controls to conduct 3 million task group analyses. Using this null data with different experimental designs, we estimate the incidence of significant results. In theory, we should find 5% false positives (for a significance threshold of 5%), but instead we found that the most common software packages for fMRI analysis (SPM, FSL, AFNI) can result in false-positive rates of up to 70%. These results question the validity of a number of fMRI studies and may have a large impact on the interpretation of weakly significant neuroimaging results.
ROLLING ROCK
Salmofan measuring salmon color
The data may not contain the answer.
And, if you torture the data long enough, it will tell you anything.
John W. Tukey
Analyzing human behavior isn't rocket science.
It's harder than rocket science.
For a successful technology, reality must take precedence over public relations, for Nature cannot be fooled.
Richard Feynman
The first principle is that you must not fool yourself - and you are the easiest person to fool
Richard Feynman
DATA ANALYSIS
confirmatory unhacked
vs.
exploratory detective work
Random error is but 1 of 20 major threats to learning from data. But random error receives disproportionate attention because it is the only threat math-modeled.
You can have your own point of view, but not your own facts.
Daniel Patrick Moynihan
DATA FORENSICS
The Quartz guide to bad data
An exhaustive reference to problems seen in real-world data along with suggestions on how to resolve them.
https://github.com/Quartz/bad-data-guide
People and institutions cannot keep their own score. Metrics become targets, and are gamed, undermined, corrupted, misreported, fudged.
LEARNING ABOUT YOUR DATA
To learn about a process as a whole, directly observe how the original measurements are made in the field.
Numbers on the screen are representations of the real world.
Look at the real world, not just representations.
Walk around what you want to learn about.
In doing creative work, do not start your day with addictive time-vampires such as The New York Times, email, Twitter.
All scatter eye and mind, produce diverting vague anxiety, clutter short-term memory.
Instead begin right away with your work.
Many creative workers have independently discovered this principle.
THE LAST INTERFACE = THE THINKING EYE
How do you I they know that?
To see with fresh different eyes: stay in the optical experience, forget the name of what one sees, defamiliarize, see with laughing eyes, construct parodies, shut up and look, see with an open mind but not an empty head, see to learn not confirm.
Unnecessary noise is the most cruel absence of care which can be inflicted on sick or well.
Florence Nightingale
To loot is to look for what is important for one's work, not letting the local discipline tell you what is important or relevant.
Looting, focusing, curiosity is this: to get so specific about something that it becomes something other than what it ordinarily is.

(For video of this talk, search *tufte microsoft keynote machine learning AI*)

CASE STUDY: MAKING IMPORTANT DECISIONS IN SENIOR FACULTY MEETINGS

The senior departmental faculty meets to decide to promote a candidate to a tenured full professorship (a lifetime job, rarely revocable). Such decisions are big bets on the future with difficult trade-offs: the possibility a better candidate might be available somewhere, and the opportunity costs of a tenured position, which counts as *two* full-time budgetary equivalents, and so the department could hire *two new junior faculty instead of giving one person tenure.*

In some departments, the tenure decision meeting begins with a reading period (senior faculty are fast readers). Each professor receives copies of the candidate's resumé, external letters of recommendation from other scholars, teaching evaluations, and instructions from some Deans about the position. Two sets of the candidate's publications are available, both in the meeting and earlier. A reading period changes the character and flow of the meeting. Colleagues are less likely to lecture one another since everyone has read the material. Also, since the external letters of recommendation are available to everyone, cherry picking the letters is reduced because everyone at the meeting has the letters in hand and has read them.

The reading period may lead to smarter/shorter meetings and greater civility. In collegial departments, I enjoyed faculty meetings (!), when colleagues discussed excellent work and celebrated an outstanding candidate. On a difficult day – a marginal candidate approved by a close vote (ending the candidate's case, since close votes are not good enough for the higher-ups to approve the appointment); a narrow specialized candidate ("among the best in their field," consisting of 20 researchers worldwide), factional divisions, conflicts of interest, divergent priorities – at least decisions are somewhat more evidence-based as a result of reading documents. At faculty meetings making important decisions, I deliberately sat next to a colleague likely to have views different from my own, since it is difficult to be uncivil with an adjacent colleague compared to one across the table.

REASONING IN DECISION MEETINGS: BEWARE OF ALL-PURPOSE ARGUMENTS

Opportunity Cost arguments are powerful and relevant, but may be used indiscriminately to reject *any* candidate or policy by comparisons with imaginary or unavailable competitors.

Dangerous Precedent arguments claim that we shouldn't do the obviously right thing now, for fear that we will be asked to do the right thing again in some hypothetical future. Thus every action "which is not customary, either is wrong, or, if it is right, is a dangerous precedent. It follows that nothing should ever be done for the first time."
(F.M. Cornford, *Microcosmographicia Academica*, 1908)

Unintended Consequences arguments divert attention from specific *intended consequences*. Unintended Consequences arguments are a mash-up of Opportunity Costs and Dangerous Precedent arguments – too often truisms, vague generalities about the consequences of unknown unknowns. But these truisms can quickly turn toward truth, becoming credible/relevant/effective when *accompanied by specific evidence concerning the issues at hand.*

SERIOUS NONFICTION MEETINGS SHOULD BEGIN WITH SILENT READING PERIOD OF A SUBSTANTIAL DOCUMENT USING PARAGRAPHS AND SENTENCES. NO BULLETS LISTS, NO SLIDE DECKS, NO REHEARSALS, NO PRESENTATIONS.

Nonfiction presentations should begin with a silent reading of a document, 2 to 8 pages long, written in sentences, with images and data displays. Documents model information better than slides. Sentences are smarter than bullet lists. Give your audience a document as they arrive, saying 'Read this, then we'll talk about it.' Meetings with multiple topics imply several silent reading periods. Audience members read 2 or 3 times faster than presenters can talk. Everyone reads with their own eyes, their own pace, their own sense of what is relevant. In slide presentations, viewers have no control over pace and sequence as presenters click through decks – while viewers sit in the dark waiting for diamonds in the swamp.

Presenters, you have not lost your contribution, after all, you prepared the document. Study hall is wonderful for presenters: people showed up, they're all reading your stuff, they're not looking at you and innumerable slides. If someone peeks at their email instead of reading, gently glare at them; the purpose of gathering together is total concentration on the content at hand. This method – document, study hall, discussion – comes with a guarantee: meetings will be smarter, more efficient, shorter. None ever wished them longer.

Slide decks, however, are much easier to prepare than documents, which is why presenters and teachers like PowerPoint. But presenter/teacher convenience in preparing decks harms the most important things in a meeting – content and audience! Optimizing presenter convenience is selfish, lazy, and worst of all, replaces thinking and content engagement:

Steve Jobs: *"I hate the way people use slide presentations instead of thinking. People confront problems by creating presentations. I want them to engage, to hash things out at the table, rather than show a bunch of slides. People who know what they are talking about don't need PowerPoint."*

Jeff Bezos on Amazon meetings: *"We have study hall at the beginning of our meetings. Staff meetings at Amazon begin with 30 minutes of silent reading. The traditional corporate meeting starts with a presentation. Somebody gets up in the front of the room and presents a PowerPoint presentation, some type of slide show. In our view you get very little information, you get bullet points. This is easy for the presenter, but difficult for the audience. And so instead, all our meetings are structured around a 6 page narrative memo."*

Even Steve Ballmer, then Microsoft CEO, pointed out big inefficiencies in PP presentations: *"The mode of Microsoft meetings used to be this: You come with a slide deck presentation. You deliver the presentation. You take a long and winding road. You take the listener through your path of discovery and exploration, and you arrive at a conclusion. That's kind of the way I used to like to do it, and the way Bill Gates used to kind of like to do it. I decided that's not what I want to do anymore. I don't think it's productive. I don't think it's efficient. I get impatient. Most meetings nowadays, you send me the materials and I read them in advance. And I can come in and say: 'I've got the following four questions. Please don't present the deck.'* [!] *If I have questions about the long and winding road and the data evidence, I can ask them."*

AMAZON: SILENT READING OF A DOCUMENT, THEN ANALYSIS AND DISCUSSION, AND NO PP

Colin Bryar and Bill Carr, *Working Backwards: Insights, Stories, and Secrets from Inside Amazon*, 2021, 79-97, edited.

"If you were to ask recently hired Amazon employees about what has surprised them most at the company, one response would top the list: 'The eerie silence in the first 20 minutes of many meetings.' After a brief exchange of greetings and chitchat, everyone sits at the table, and the room goes completely silent. Why? A six-page document that everyone reads before discussion begins. Amazon relies far more on written words to develop and communicate ideas than most companies, and this difference makes for a huge competitive advantage.

The End of PowerPoint: Our Inspiration [Edward Tufte]

One of my roles as Jeff's shadow in the early days was to manage the agenda of weekly meetings, which took place Tuesdays and typically ran for four hours. 80% of the time was focused on how the company was making progress toward achieving goals. The meeting was expensive: preparation and attendance consumed at least half a day each week for the top leaders. Given the decisions made in the meeting, the stakes were high.

In those early days, deep dives would begin with a presentation by the team on the status of their work. Typically, this involved an oral presentation backed up by PowerPoint slides. Too often, the presentations did not serve a purpose. The format made it difficult to evaluate actual progress. The deep dives were frustrating, inefficient, error prone.

Jeff and I often discussed ways to improve the S-Team meetings. After a difficult presentation in early 2004, on a business flight we read and discussed an essay "The Cognitive Style of PowerPoint: Pitching Out Corrupts Within," by Edward Tufte, a Yale professor who is an authority on information visualization. Tufte identified in one sentence our problem:

'As analysis becomes more causal, multivariate, comparative, evidence based, and resolution-intense, the more damaging the bullet list becomes.'"

"Tufte's description fit our discussions at the meetings: complex, interconnected, requiring plenty of information to explore. Such analysis is not well served by a progression of slides that makes it difficult to refer one idea to another. The Amazon audience of tightly scheduled, experienced executives was eager to get to the heart of the matter. They would pepper the presenter with questions and push to get to conclusions. Sometimes the questions did not serve to clarify a point or move the presentation along but would instead lead the entire group away from the main argument. Some questions might be premature and would be answered in a later slide.
In his essay, Tufte proposed a solution:

"For serious presentations, replace PowerPoint slides with paper handouts showing words, numbers, data graphics, images together. High-resolution handouts allow viewers to contextualize, compare, narrate, and recast evidence. In contrast, data-thin, forgetful displays tend to make audiences stupid and passive, and also to diminish the credibility of the presenter."

Tufte's wise advice on how to get started:

"Making this transition in large organizations requires a straightforward executive order: 'From now on your presentation software is Microsoft Word, not PowerPoint. Get used to it.'"

That is essentially what we did. On June 9, 2004, the members of the S-Team received an email with this subject line: "No PowerPoint presentations from now on at S-Team." This message was simple, direct, earthshaking: from that day on team members were required to write short narratives describing their ideas. PowerPoint was banned.

A narrative document anticipates objections, concerns, alternative points of view, questions, common misunderstandings. Edward Tufte bluntly sums up the benefits of narratives over PowerPoint:

'PowerPoint becomes ugly and inaccurate because our thoughts are foolish, but the slovenliness of PowerPoint makes it easier for us to have foolish thoughts.'

How to Conduct a Meeting in This New Format

When everyone has read a document, some presenters start by saying, "Let me walk you through the document." The attendees have already walked themselves through the argument. Audience members now ask questions, seek clarification, probe intentions, offer insights, suggest refinements or alternatives. The key goal is to seek the truth. During the discussion, notes are taken, by someone knowledgeable about the subject who is not the primary presenter. If I don't see anyone taking notes, I would politely ask who is going to do so.

Ideas Matter Most

The entire team can contribute to crafting, reviewing, revising the narrative. Decisions draw from ideas, not individual performance skills. The time spent upon crafting gorgeous, graphically elegant slide presentations can now be used for more important things. What matters is found on the page. Anyone can edit or make comments on the document, and they are easily shared in the cloud. The document serves as its own record. Narratives are designed to increase the quantity and quality of effective communication – by an order of magnitude over traditional methods. The narrative team writes its first draft, circulates, reviews, iterates, repeats, and then finally takes the vulnerable step of saying to their management and peers, 'Here's our best effort. Tell us where we fell short.'

This model imposes duties and expectations upon the audience as well. They must evaluate the idea, not the team or the pitch. The work product of the meeting is a joint effort of the presenters and their audience. Silence during the discussion is the equivalent of agreement with what is presented, but it carries the same weight as a full-blown critique. Presenters and audiences become linked to the subsequent success or failure of the initiative and business analysis. When looking at Amazon's big wins, every major success has gone through multiple narrative reviews, with contributions from the presentors and their audience."

Colin Bryar and Bill Carr, *Working Backwards: Insights, Stories, and Secrets from Inside Amazon*, 2021, 79-97, edited.

REMODELING DOCTOR-PATIENT COMMUNICATION: PATIENTS PRESENT A LIST OF CURRENT MEDICAL CONCERNS TO THEIR DOCTOR*

Some 50 years ago, studies of doctor-patient interactions reported that doctors, at the beginning of an appointment, interrupted patients who were talking after 18 seconds. Recent replications reported 11 to 22 seconds prior to interruption. At the beginning, the state of the doctor's knowledge is uncertain, although many may have studied the patient record. Medical records, however, belong to, and are copyrighted by a particular medical clinic, making sharing difficult. This leads to expensive duplicate tests. Veterinarians do much better at sharing medical records.

In advance of any medical event, patients should provide a list of concerns, symptoms, questions. The list, recent and relevant to the current appointment, can be prepared by the patient or someone who knows the patient. The list might also report speculations by the patient: "Pain in my right side might be gall-bladder, like one of my siblings." The patient makes several copies of the list, brings them to appointments, hands the list to the doctor. (I observed a doctor read the list out loud as the patient interrupted.) In telemedicine, patients should send their list promptly at the moment the meeting begins.

Doctors, who did not get to be a doctor by being a slow reader, can read several times faster than patients can talk. After handing the list over, the patient should look pointedly at their own copy, hinting it is time to start reading, saying "Here it is, I have 5 issues." If the doctor just glances at the list and says "What's up?"– then the patient should just calmly read the list aloud! The list of medical concerns gets everything the patient initially has to say out on the table, without interruption. As the appointment continues, lists set an agenda for the allocation of time. Doctors are taught to orally elicit all of the patient's agenda early in appointments and set priorities.

Lists also makes sure that patients and doctors do not abandon lower-level issues. Each item on the list is, in effect, checked off as the appointment moves along. The patient might ostentatiously check off the first point on the list after it is discussed to indicate this list is what we're going to march through. A list read early in the appointment avoids one of the things that bothers doctors the most – patients saying "By the way, one more thing, I have chest pain," as the visit is concluding. This is usually the main problem patients are worried about, and is often something serious.

Patients should bring several copies of the list, since they may interact with doctors, nurses, medical students, techs. All get the list. For example, a doctor sometimes has a medical student in training who handles the initial discussion and then goes off to describe the situation to the doctor, who shows up later. Both student and doctor should receive copies of the list. Multiple copies are particularly effective in the emergency room (if the patient or a friend can prepare a list), since the patient will surely see many medical staff members in the emergency room. If the patient is then transferred into the hospital for observation, a refreshed and updated list is helpful. Otherwise, just use the original list.

---

* This section on medical problem lists is written by Dr. David S. Smith (Yale Internal Medicine, author of *Field Guide to Bedside Diagnosis*) and by E.T.

The Medical Concerns List enhances the efficiency, accuracy, and resolution of doctor-patient interactions. It also helps reduce socially or situationally-determined answers to the doctor's questions that might be embarrassing to the patient. This list goes into the patient record and also assists the doctor in preparing notes for the patient record. Don't send pre-meeting list (doctors are busy with patients already), instead provide the list at the exact moment of need, the beginning of the appointment.

Prepare your list in the serenity and privacy of your home

In your list, say what's going on, when it started, how often particular symptoms occur. Precise relevant details will help. If the problems involve occasional incidents, make a short video. If relevant, mention your family history. Review your list with someone who knows you well; they may see and recall things that you don't. Review your recent medical lists and adventures. Do your homework, rehearse for your appointment.

If you search the web about your health concerns, go only to credible websites. The Mayo Clinic Symptoms Guide is best. Avoid predatory websites (highly ranked by Google advertising), avoid web charlatans and crackpots (if their grandiose claims were true, where is their Nobel Prize in Medicine?), avoid drug company ads and other medical product pitches – the result of targeted advertising from continuous tracking of your web activities. Google knows all about you.

Anticipate the exact moments of need for your list

Your list will serve you well when you're sitting on an examination table in a paper dress, stressed out, and it's hard to remember everything. No problem, you've prepared your document at home in advance. Before surgery, mention your concerns to the anesthesiologist – "Sore throat for a week after last anesthesia, my pipes are narrow" "Have had PVCs for years." Bring a list to every encounter. For complex or serious problems, identify and go to higher ranked medical systems – and this independent replication may well yield confirming/better/divergent results.

After appointment

As a result of the Cures Act (U.S.), patients now are able to access their electronic health records and can see their doctor's analysis and instructions. At the end of an appointment, some medical centers provide a paper copy, since not all patients can manage electronic records. Patients can then consider their next steps, checking up on proposed treatments or tests by looking at the *Choosing Wisely* website and other reports of low-value treatments. Patients thus might avoid getting caught up in cascades of over-diagnosis, over-testing, duplicate tests, over-treatment. In the U.S., patients are part of business plans seeking to maximize profits. For example, investment banks now own many Emergency Departments and ambulance services, notorious for high prices, surprise billing, and automated legal pursuit of unpaid bills via bankruptcy courts.

Conclusion

Medical Concerns Lists (MCLs) can help patients communicate with the medical world more effectively and precisely. These gains require no increase in size of vast medical business bureaucracies.

SELF-PUBLISHING OF BOOKS: REMODELING EVERYTHING

Early in my career as a professor, I wrote books on public policy and data analysis. Two of the book covers showed my artworks, covert efforts that started as a child and got serious (a week per month) when I started teaching. The books were good, the paintings not. Moving into a house with a yard led to outdoor concrete sculptures, and I greatly enjoyed the physicality of artwork and design. Writing back then was physical as well – an IBM Selectric typewriter with its good keyboard feel, cutting up and taping down layers of text edits. I served on editorial boards of journals, and saw first-hand intense writing, editing, and publishing to deadlines as a newsroom consultant on elections, working with news reporters and editors. I was an insider, but within myself always an outsider, a stranger observing, questioning, learning.

Six years into my careers, I started collecting material for what became *The Visual Display of Quantitative Information*. I gave a talk (with 16 page handout!) in a 1975 conference on data graphics. By 1981, the book was apparently ready for publication, but too much was based on U.S. graphics since 1950. It took another 18 months of work to demonstrate that my ideas applied full-time everywhere by bringing in scientific and Asian data graphics done for hundreds of years. Ever since, the same idea, forever knowledge.

A publisher was interested but planned to print only 2,000 copies and charge a very high price, contrary to my hopes for a wide readership. I sought to design the book so as to make it *self-exemplifying* – that is, the physical object itself would reflect the intellectual principles advanced in the book. Publishers seemed appalled at the prospect that an author might govern design. Consequently I decided to self-publish the book. This required a first-rate book designer, a lot of money (at least for a young professor), and a large garage. I found Howard Gralla who had designed many museum catalogs with great care and craft. He was willing to work closely with this difficult author who was filled with all sorts of opinions about design and typography. We spent the summer in his studio laying out the book, page by page. We integrated graphics into the text, sometimes in the middle of sentences, eliminating the usual segregation of text and image – one of the ideas *Visual Display* advocated. To finance the book I took out another mortgage (at 18% interest) on my home. The bank officer said this was the second most unusual loan that she had ever made; first place belonged to a loan for a circus to buy an elephant!

My view on self-publishing was to go all out, to make the best and most elegant and wonderful book possible, without compromise. Otherwise, why do it? The next 4 books were financed by the previous books. I have never written a grant application.

Authors considering self-publishing: Publish several books with real publishers first. All authors should keep copyrights to their books. Most self-publishers wind up with a basement of mildewing books.

REMODELING COLLEGE TEACHING: ONE-DAY COURSE WITH 5 BOOKS AS HANDOUTS, GO OUT AND TRAVEL TO THE STUDENTS (OF ALL AGES). NO NEED FOR COLLEGE ADMIN BUREAUCRACY AND AGE SEGREGATION

For 27 years I taught a live one-day course "Presenting Data and Information" for 923 days to 328,001 students in hotel ballrooms. Then, during the pandemic, the course went online: people sign-up, every student is shipped copies of 5 books and a password for a 4-hour video lecture closely integrated with the video. Both the live and online courses begin with a reading assignment, followed by short reading periods during the course. My books served as takeaways – more precise, thorough, and extensive than a class. The courses are a product of writing, designing, and self-publishing my books via my Graphics Press. I like hotel ballroom venues compared to college classrooms: no age discrimination as in universities (why teach only 18 to 22 year olds?), happily resulting in an interesting, lively, and diverse audience – heart surgeons, very young prodigies, epidemiologists, and a chicken farmer (one told me he took the course because chickens have data).

In small groups, questions involve engagement and discussion. In big groups, not so much. I enjoyed questions in big groups because they provided prompts to speak informally and vividly. However, people who ask questions in large groups tend toward mini-speeches not questions. But consider the priorities of audiences: *first, the person they would like most to talk are themselves; second, they would like to hear the presenter; and third, the people they least like to hear talk are their fellow audience members*. Thus the great compromise: the presenter will talk. In doing Q and A the same questions often show up, so I provided answers to common questions in my lectures and handouts.

My view is to do the best I can and put it out there, largely indifferent to what people might think.

My course material changes ~15% per year, as each book currently in progress becomes part of the course years before finally published. Teaching reveals incoherencies and mistakes in new material, leading to refinements or even throwing stuff out from the manuscript for the upcoming book. A good way to learn about something is to teach it.

9 REMODELING THE BACK-MATTER IN BOOKS
A VISUAL INDEX, A QUILT OF SOURCES AND IMAGES

Not all that many readers go to the back matter and look up the source for a single sentence. But the *back matter can become multimodal paragraphs*, revealing a history of content and sources. And images and illustrations from the book in the back matter create a lovely visual/verbal summary quilt of the entire book, enjoyed by all.

Some images are edited and redrawn (indicated in citations) to repair battered originals, make color separations, improve design. Primary sources, themes for my variations, are indicated below. When relevant, citations are placed directly in the text for the convenience of readers and to assist understanding the content. 60 uncredited photographs were taken by ET.

COVERS

Edward Tufte, 'Lepton g-2 10th order Feynman diagrams and integrals,' describing subatomic particles, stainless steel artwork, 2018-2000.

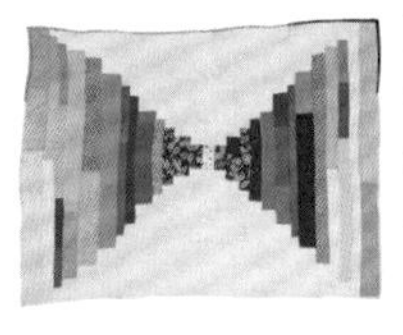

Loretta Pettway, *Log Cabin, Single Block Courthouse Steps* variation, local name *Bricklayer*, a cotton Gees Bend quilt, 70 × 84 inches, 1959 © 2019 Loretta Pettway/Artists Rights Society (ARS) New York

ET and artworks, Hogpen Hill Farms, May 2020, photo by Cynthia Bill

INTRODUCTION

4 **The most dangerous phrase in the language** Grace Hopper in Esther Surden, 'Privacy Laws May Usher in Defensive DP,' *Computerworld*, 10, 26 January 1976, 9

5 **Good ideas are a dime a dozen for a smart person** Craig Venter, *A Life Decoded: My Genome, My Life*, 2007, 129

1 MEANING AND SPACE, REMODELING CONVENTIONS

6 **Music is the space between the notes**
Miles Davis, Claude Debussy, Ben Jonson, et al

6 **The horizon is not a line, it is a space**
Joanne K. Cheung, Twitter, 9 January 2019

6 **A vessel is useful only through its emptiness**
Lao Tse, in György Kepes, *The Language of Vision*, 1948, 32

6 **This subject — turning the line in a poem — is one that every poet deals with** Mary Oliver, *A Poetry Handbook*, 1994, 35-36

6 **Vacuum, in modern physics, is what you get when you remove everything you can** Frank Wilczek, 'How Feynman Diagrams Almost Saved Space,' *Quanta Magazine*, 5 July 2016, edited

6 **There is no there there** Gertrude Stein, *Everybody's Autobiography*, 1937, on visiting her hometown, Oakland, California

6 **Make sure your code 'does nothing'** Brian W. Kernighan and P. J. Plauger, *The Elements of Programming Style*, 1978, 111

6 **Nothing that is not there and the nothing that is**
Wallace Stevens, 'The Snow Man,' *Harmonium*, 1923

6 **What are you rebelling against? What do you got?**
Dialogue between Mildred and Johnny, 1953 film *The Wild One*

7 **Rationalists, wearing square hats** Wallace Stevens, 'Six Significant Landscapes, VI,' *Harmonium*, 1923

7 Andrei Severny photo of ET, 2018, Guggenheim Museum

8 **Research on horizontal gene transfer**
David Quammen, *The Tangled Tree: A Radical New History of Life*, 2018; Carl Zimmer, *She Has Her Mother's Laugh: The Powers, Perversions, and Potential of Heredity*, 2018

8 **Organizations which design systems** Melvin E. Conway, 'How do Committees Invent?,' *Datamation* 14, 1968, 28-31, edited

8 **Der Dogg und Der Überdogg** Daniel Pinkwater (mocking Konrad Lorenz, *Man Meets Dog*), *Uncle Boris in the Yukon and Other Shaggy Dog Stories*, 2001, 76-77

8 **stylized facts** Robert Solow, *Growth Theory*, 1970, 2

8 Erle Loran, *Cézanne's Composition: Analysis of His Form with Diagrams and Photographs of His Motifs*, 1943, 76-77; Paul Cézanne, *La table de cuisine*, 1888

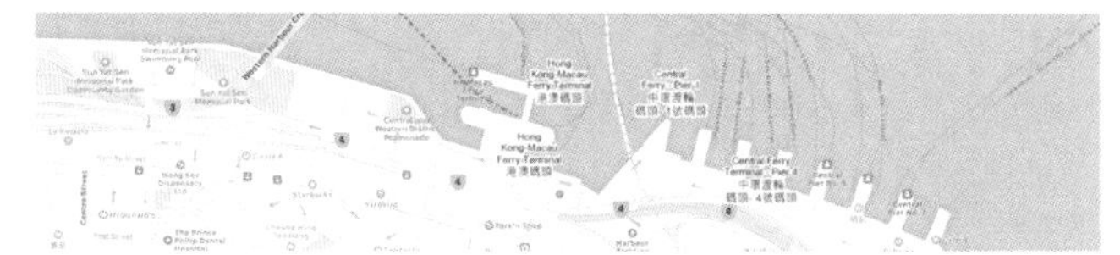

9 Hong Kong waterfront, multiple map collage, July 2020

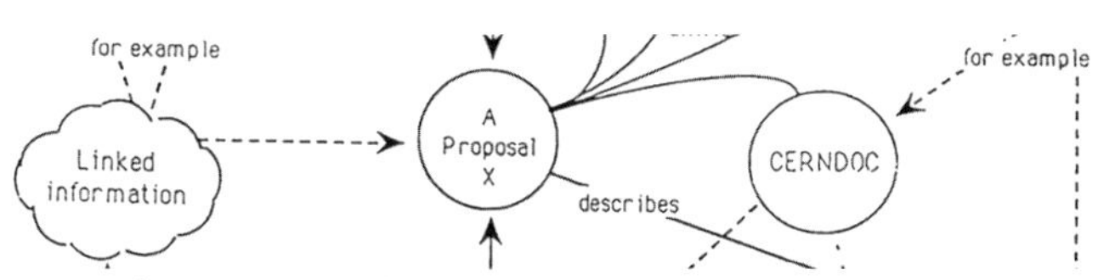

11 The problem with trees: Many systems are organised hierarchically Tim Berners-Lee, 'Information Management: A Proposal,' CERN, March 1989, edited

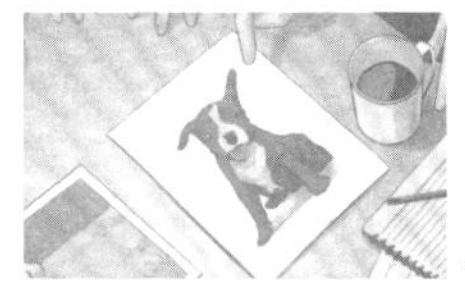

12 Exactly how we collaborated mattered Ken Kocienda, *Creative Selection: Inside Apple's Design Process During the Golden Age of Steve Jobs*, 2018, 153 – 156, edited

14 Knuth-Morris-Pratt algorithm in C, Christian and Baud Lovis, 'Fast Exact String Pattern-Matching Algorithms,' *Journal of the American Medical Informatics Association*, 7, 2000, 378-391

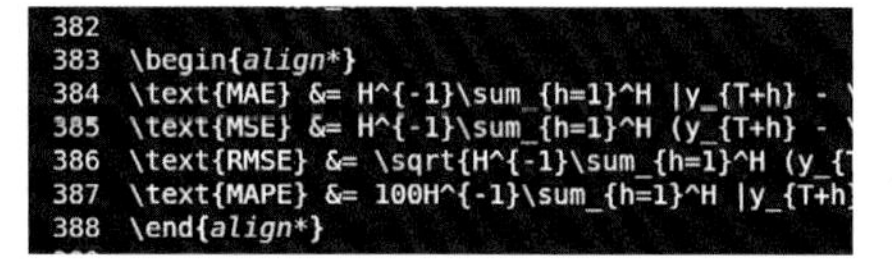

14 Rob Hyndman, lecture slide, writing code in R with Sublime

15 More than meter, more than rhyme James Longenback, *The Art of the Poetic Line*, 2008, xi

15 Historically, line is the characteristic unit Helen Vendler, *The Breaking of Style*, 1995, 71, on the poetry of Jorie Graham

15 Whether end-stopped or enjambed
Edward Hirsch, 'Line,' *A Poet's Glossary*, 2014, 348-349

15 The seasons are no longer what they once were
John Ashbery, 'Syringa,' *Houseboat Days: Poems*, 1977

15 April is the cruellest month T. S. Eliot, *The Waste Land*, 1922

16 Quantum superposition, Roger Penrose, *Road to Reality: Complete Guide to the Laws of the Universe*, 2004, 631, edited

16 Michael Borinsky, 'Algebraic lattices in QFT renormalization,' *Letters in Mathematical Physics* 106, 2016, 879, edited

17 Music is the space between the notes
Miles Davis, Claude Debussy, Ben Jonson, et al

17 Why Most Published Research Findings Are False,
John P.A. Ioannidis, *PLOS Medicine*, 2005

17 When a formula is too long T.W. Chaundy, P.R. Barrett, Charles Batey, *The Printing of Mathematics*, 1954, 38, edited

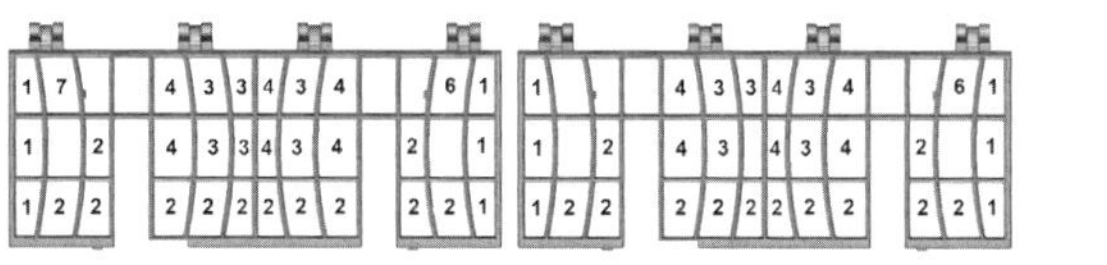

18 Dishwasher manual. Robert Bosch GmbH, 2015

19 Text and film stills, *Empathy: Exploring Human Connection*, Cleveland Clinic Media Production, 2013, edited

21 Electronic Health Records Atul Gawande, 'Why Doctors Hate Their Computers,' *The New Yorker*, 12 November 2018

21 It's your body Eric Topol, 8 October 2017, Twitter

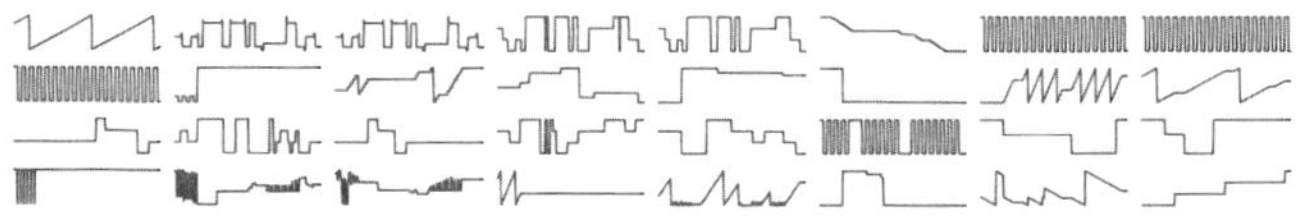

22 Michael Fogleman, 'Ten Seconds of PAC-MAN,' January 2018

23 It's open-ended, Commemorating women at Yale Maya Lin, interview, Bill Moyers, PBS, 'Becoming American: The Chinese Experience,' 2003.
Maya Lin, *The Women's Table*, Yale University; photos *Yale Alumni Magazine*, 2014, Mark Alden Branch, 4 January 2019

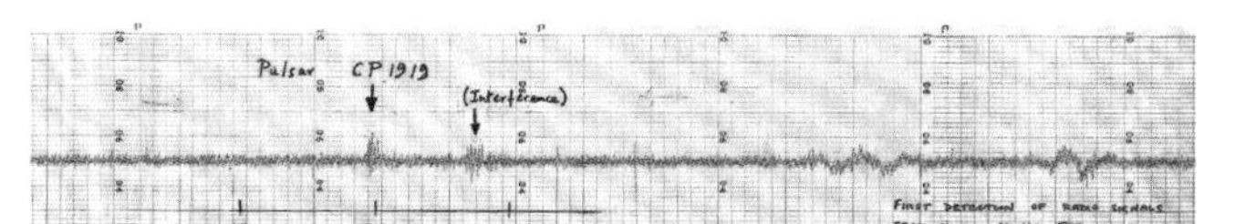

24 Jocelyn Bell Burnell, a postgraduate student at Cambridge, detected radio pulsars 6 August 1967, photo Cambridge University

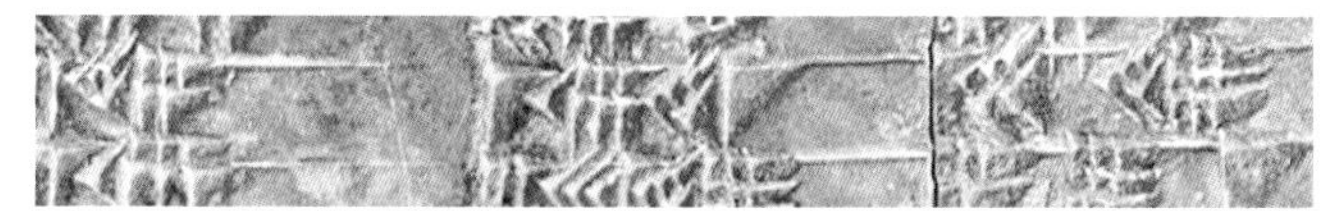

24 Babylonian clay tablet of Pythagorean triples, ~1800 BCE, Plimpton Collection 322, Columbia University; photo Christine Proust, Nicholas Wade, 'An Exhibition That Gets to Root of Sumerian Math,' *New York Times*, 22 November 2010

24 Laboratory notebook of Caltech graduate student Linus Pauling, x-ray diffraction data, Pauling Notebook 2-5, November 1922, Oregon State University, Ava Helen and Linus Pauling Papers

25 Étienne-Jules Marey, with a chronophotographic gun, captured 12 consecutive frames per second, Études pratiques sur la marche de l'homme: expériences faites à la station physiologique du Parc des Princes,' *La Nature* 608, 1885; Marey, *Le Mouvement*, 1894

25 Stroke sequence for Chinese character, chinese-tools.com

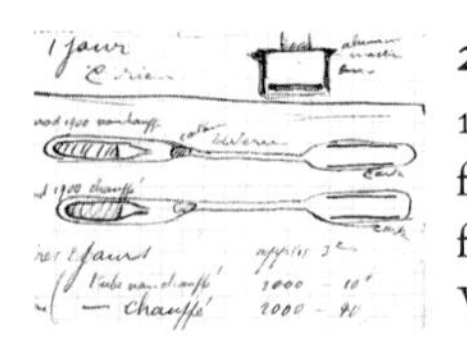

25 Holograph notebook of Marie Curie from 1899-1902, dated 19 January, containing notes from experiments on extracting radium from pitchblende with drawings of apparatus, Wellcome Library, Wellcome images L0021265.

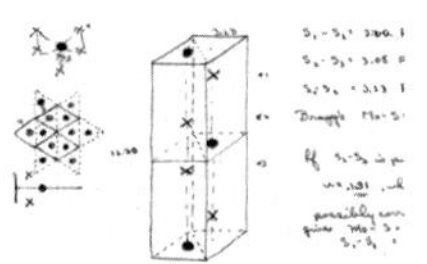

25 Linus Pauling, laboratory notebook with x-ray diffraction data, Notebook 2-41, November 1922, Oregon State University Libraries, Ava Helen and Linus Pauling Papers

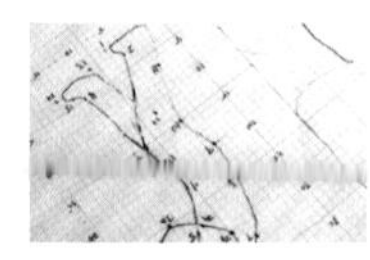

25 Susan D'Agostino, 'The Computer Scientist Who Can't Stop Telling Stories.' *Quanta Magazine*, 16 April 2020, photograph by Jill Knuth

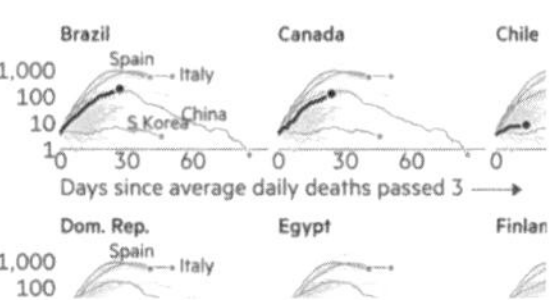

26 John Burn-Murdoch, *Financial Times*, data from European Centre for Disease Prevention and Control, 19 April 2020

27 Yehudi Menuhin's marked copy of J. S. Bach's *Solo Violin Sonata No. 2 in A minor*, 1720, Foyle Menuhin Archive

28 Photography is all right, if you don't mind looking at the world
David Hockney, in Lawrence Weschler, Hockney's *Cameraworks*, 1984

28 Every photograph here is taken close to something
David Hockney, video interview, J. Paul Getty Museum

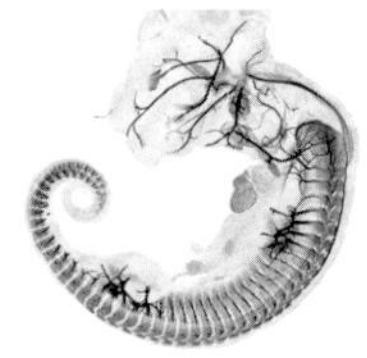

28 Daniel Smith-Paredes and Bhart-Anjan Bhullar, Madagascar ground gecko embryo after 12 days of incubation in the egg, Jim Shelton, 'Yale lab earns a top prize for its latest scientific visualization,' *Yale News*, 7 November 2018, edited

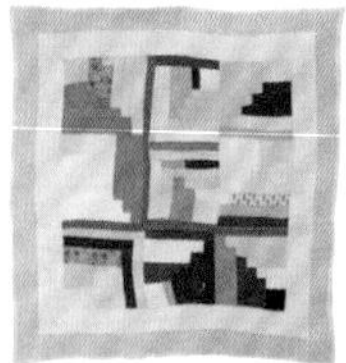

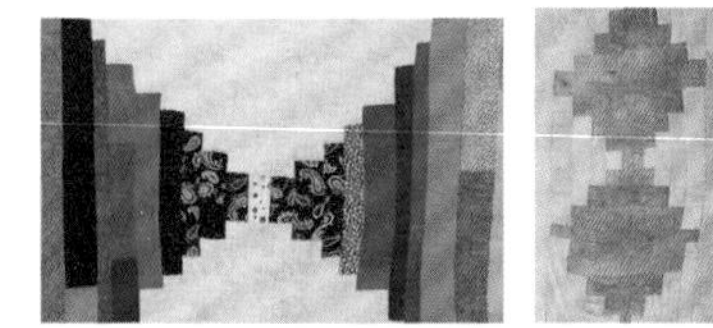

29 Jessie T. Pettway, *Log Cabin*, corduroy, twill, printed shirt, dress material, seersucker, 1960, Souls Grown Deep Collection; Loretta Pettway, *Log Cabin, single block Courthouse Steps*, 1959, Souls Grown Deep Collection; Leola Pettway, *Log Cabin: Courthouse Steps*, 1975; Auburn University, Jule Collins Smith Museum of Fine Art, Souls Grown Deep Collection. © 2019 Jessie T. Pettway/Artists Rights Society (ARS, NY). © 2019 Loretta Pettway/Artists Rights Society (ARS, NY). © 2019 Leola Pettway/Artists Rights Society (ARS, NY)

32 *Matterhorn, Landeskarte der Schweiz*, 1347, Bundesamt für Landestopographie, 1983, scale 1:25,000

32 Google internal map of San Francisco, Alexis C. Madrigal, 'How Google builds its maps,' *The Atlantic*, 6 September 2012.

33 Hogpen Hill Farms, sculpture field, aerial photo from Google Maps, Connecticut, 2019

33 ET, *Megalith with 6 Elements*, 2013, native stone and air. Wendy MacNaughton, sketches, 2015; ET planning sketches, 2011. ET, *Continuous Silent Megaliths: A Structure of Unknown Significance*, 2012-2017, native stone and air

33 ET, *Escaping Flatland*, 2001-2008, stainless steel; Hogpen Hill Farms

33 ET, *Larkin's Twig* with visiting sheep, steel and air, 2004
ET, *I-Beam Megalith*, 2013, native stone, steel, air
ET, *Bamboo Maze*, 2011-2020, Hogpen Hill Farms, Connecticut

33 ET, *African Siberian Dancers*, series of 15+, 2016-2020, steel, Hogpen Hill Farms, derived from prehistoric rock art in Africa, Leo Frobenius and Douglas C. Fox, *Prehistoric rock pictures in Europe and Africa*, 1937

33 ET, Black Swan, 2017

33 ET, *Dancer with Calipers*, 2020

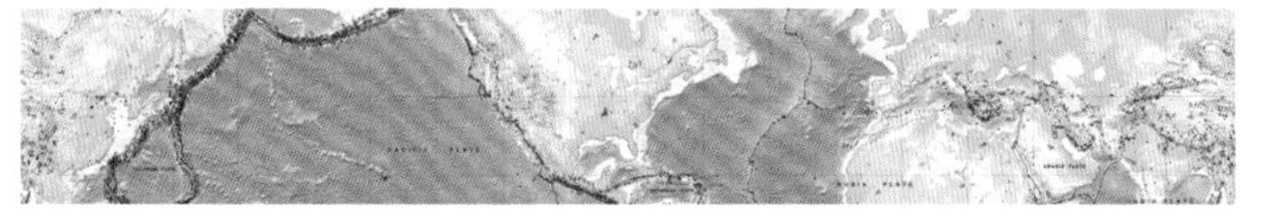

34 Tom Simkin, Will Stettner, Antonio Vellaseñor, Katherine Schindler, 'World Map of Volcanoes, Earthquakes, Impact Craters, and Plate Tectonics,' USGS, Smithsonian Institution, U.S. Naval Research Laboratory, 2006

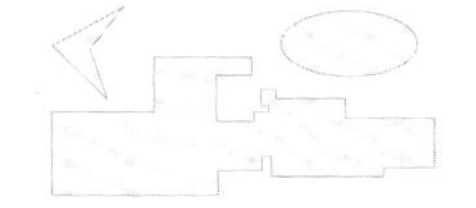

35 Planviews drawn by Pamela Pfeffer, student project, Studies in Graphic Design, Yale University, 1988

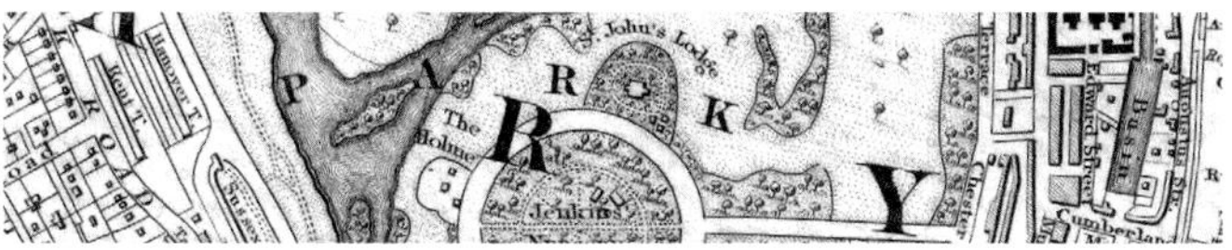

35 William Schmollinger, Regent's Park, handcolored 'Improved map of London for 1833, from Actual Survey.'

36 ET, *Larkin's Twig*, 2004, steel and air, Hogpen Hill Farms. Watercolor tone added to image by Wendy MacNaughton

37 Richard Serra, *Junction*, 2011, ET photo, 17 September 2011

38-39 Laurent de Brunhoff, press proof 1946, *Babar et ce coquin d'Arthur*, 1946, photograph, Sotheby's Paris, 'Bande dessinée,' 4 July 2012, lot 27

38-39 Haystack paintings, Claude Monet, late summer 1890-spring 1891

40 Maps show what we think we know Chris Quigg

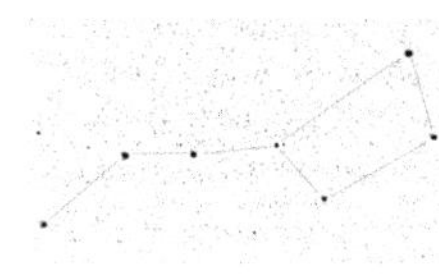

40 Fényes Lóránd, 'Big Dipper, Deep Sky,' 16 January 2016, edited

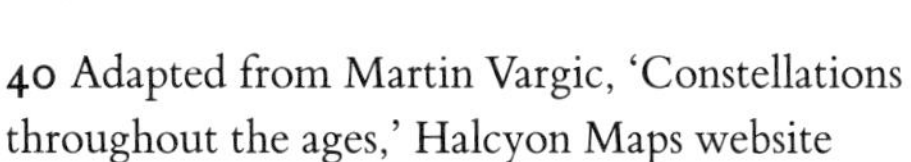

40 Adapted from Martin Vargic, 'Constellations throughout the ages,' Halcyon Maps website

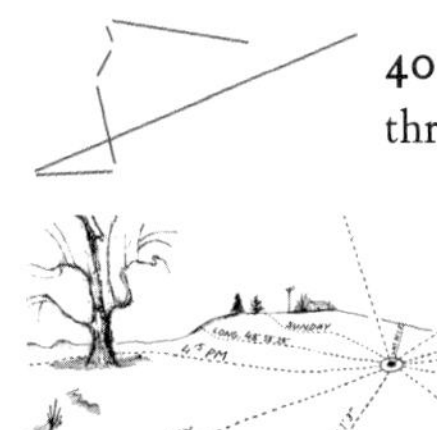

41 Saul Steinberg, *The New Yorker*, 24 February 1968, 41.

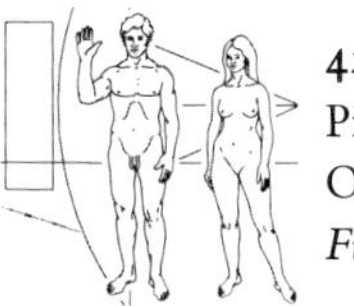

42 Pioneer plaque, gold-anodized aluminum, 1972 Pioneer 10, and 1973 Pioneer 11 spacecraft; Richard O. Rimmel, James Van Allen, Eric Burgess, *Pioneer: First to Jupiter, Saturn, and Beyond*, 1980, 248-250

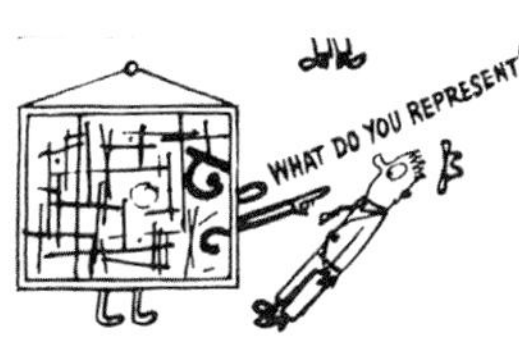

44-45 Ad Reinhardt, *How to View High (Abstract) Art*, 1946, *PM* newspaper offset lithograph, University of Kansas, Spencer Museum of Art

45 Philosophers will say that humans can never John Gray, *The Silence of Animals: On Progress and Other Modern Myths*, 2014

46 Austin Kleon, *To Have an Open Mind But Not an Empty Head, Tufte*, May 2015

46 Creativity is just connecting things Steve Jobs, in Gary Wolf, 'The Next Insanely Great Thing,' *Wired*, 1 February 1996

47 And what is it you know, once you think you know Maira Kalman, *The Principles of Uncertainty*, 2007, 129

47 What do we know that is not true? Chris Quigg, Twitter, February 2020

2 CONTENT-RESPONSIVE TYPOGRAPHY

48 Typography exists to honor content.
Robert Bringhurst, *The Elements of Typographic Style*, 1992, 17

48 There were books of all kinds Somerset Maugham, 'The Book-Bag,' 1933

ta imperfezzione dello ſtrumento, ò dell'occhio del riguardan- *Diuerſità*
te,perche ſendo la figura di Saturno coſi ,come moſtra- *nel veder*
no alle perfette viſte i perfetti ſtrumenti , doue manca tal *Saturno*
perfezzione appariſce coſi non ſi diſtinguendo perfetta- *cagionata*
mente la ſeparazione , e figura delle tre ſtelle ; ma io che mil- *da diſetto*

49 Galileo Galilei, 'Lettere di Galileo e di Marco Velsero intorno alle macchie solari,' *Istoria e dimostrazioni intorno alle macchie solari*, 1613, 25

50 Charles Darwin, *On the Origin of Species by Means of Natural Selection*, 1859 (first edition), table of contents

51 Guillaume Apollinaire, 'Il pleut' ('It rains'), typeset by M. Levé, *Sons Idées Couleurs* 12, 1916; *Calligrammes: poèmes de la paix et da la guerre*, 1918

52 Why Most Published Research Findings Are False John P.A. Ioannidis, *PLOS Medicine* 2.8, 2005

53 Vacuum, in modern physics, is what you get when you remove everything you can Frank Wilczek, 'How Feynman Diagrams Almost Saved Space,' *Quanta Magazine*, 5 July 2016

53 Hypertext is simply electronic text containing direct links, wikipedia

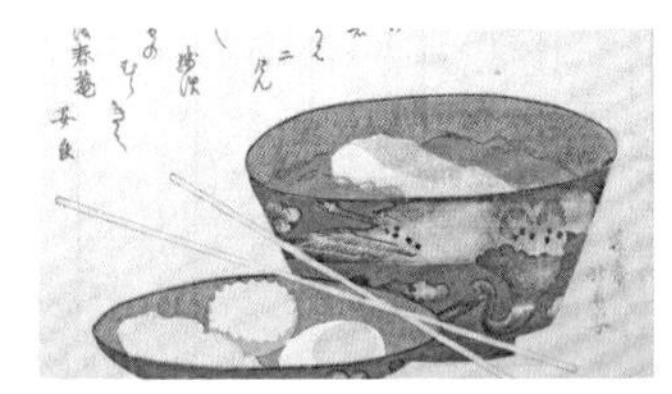

54 Teisai Hokuba, *Bowl of New Year Food*, woodblock print, 1808

54 Hieroglyph of vulture, *Neophron percnopterus*, tomb painting in Meidum in Egypt, Griffith Institute, Oxford; Henry George Fischer, *Ancient Egyptian Calligraphy*, 9, edited. Multimodel paragraph constructed by ET.

55 'Dimpling and narrowing at the toe of the anastomosis,' Siavosh Khonsari and Colleen Sintek, *Khonsari's Cardiac Surgery: Safeguards and Pitfalls in Operative Technique, 3rd edition*, 2017, 158, edited

55 **Andy Murray surgery** Instagram 29 January 2019, 'What's Wrong with this Picture?,' Association for Vascular Access; further annotation by British doctor Peter Lax, 30 January 2019

56 Saint-Genis-Pouilly area and CERN on the French-Swiss border. Swiss National Maps, Federal Office of Topography swisstopo, 2018

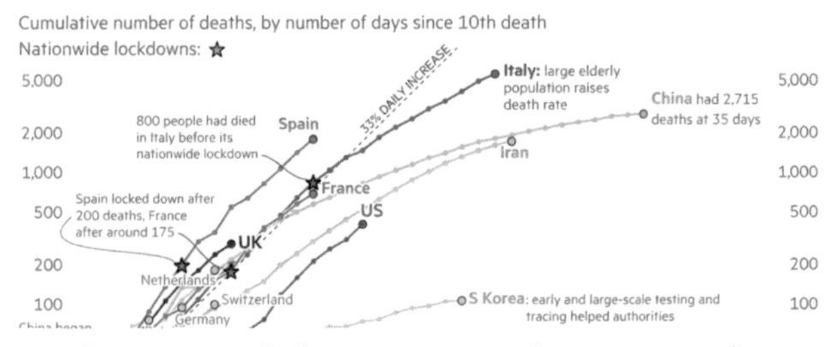

57 John Burn-Murdoch, Twitter 22 March 2020, *Financial Times*, analysis of Johns Hopkins University, CSSE; Worldometers on Coronavirus deaths

58 Elisabeth Gasteiger et al, Metabolic Pathways, U.S. DOE, 2006

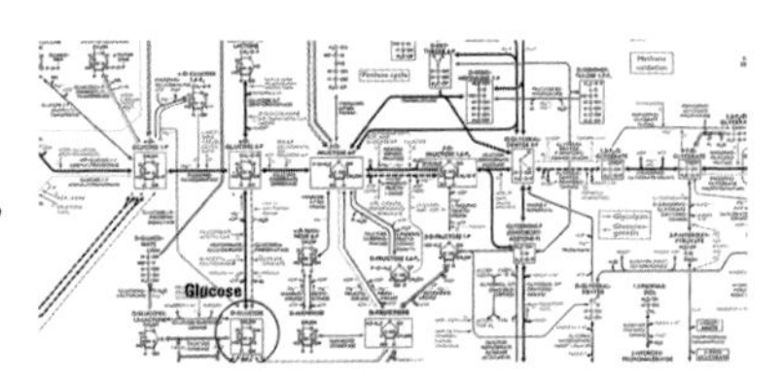

58 A striking feature of metabolism Norman R. Pace, 'Universal nature of biochemistry,' *PNAS* 98, 2001, 805 - 808

58 Blood pressure trial Jackson Wright et al, 'Randomized Trial of Intensive v. Standard BP Control,' *NEJM* 373, 2015, 2103 - 16, edited

60 Silence is the basis of music Alfred Brendel, *A Pianist's* A - Z, 2013

60 In the recording, the woodwinds play *Absolutely on Music: Conversations, Haruki Murakami with Seiji Ozawa*, translated from Japanese by Jay Rubin, 2016, 21 - 22

62 Graphical display of random behavior from Donald E. Knuth, *The Metafont Book*, 1986, 183, 251

3 GRAPHICAL SENTENCES: NOUNS AND VERBS, STRUCTURE AND FUNCTION

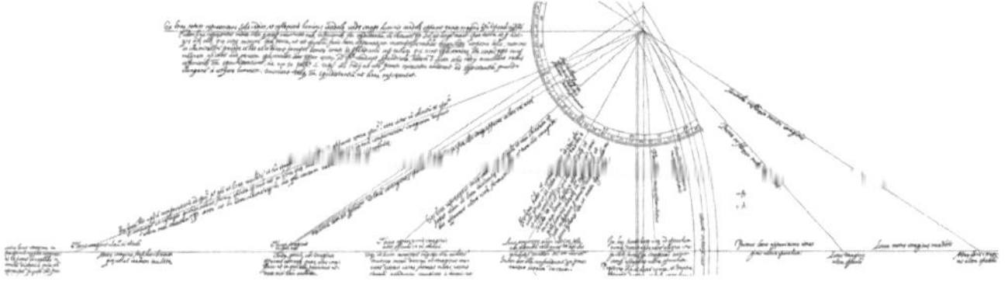

64 'Theory of the Concave Spherical Mirror,' Ettore Ausonio, ~1560; copy made by Galileo Galilei 1592 - 1601; Florence, Biblioteca Nazionale, Gal. 83, f. 4r; image in *Le Opere di Galileo Galilei*, edited by Antonio Favaro, 1907, v.3.2, 865 - 867, translated by Sven Dupré, *Galileo, the Telescope, and the Science of Optics in the XVI*, 2002

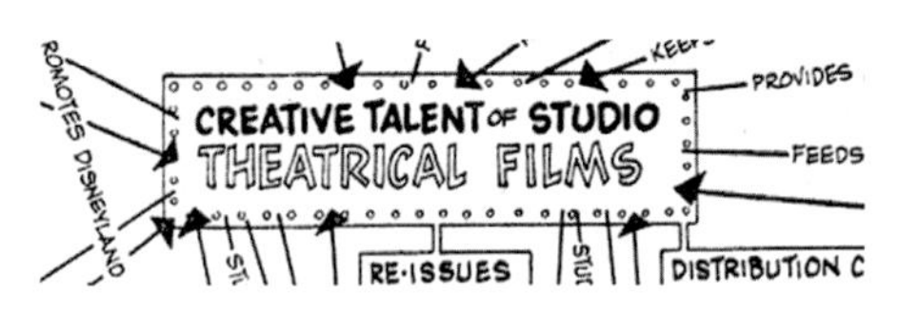

68 1957 drawing by Walt Disney, archives, The Walt Disney Family Museum

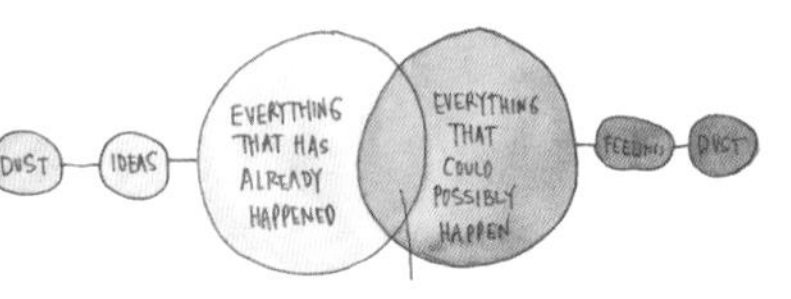

69 Wendy MacNaughton, *The Universe and Forever*, print, 2003

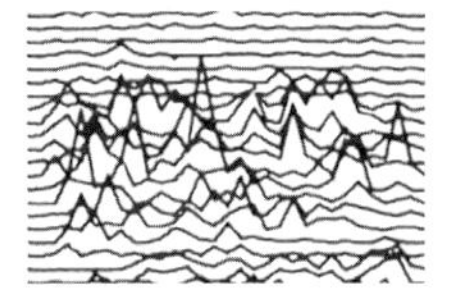

69 Timothy H. Hankins and Barney J. Rickett, 'Pulsar Signal Processing,' in Berni Alder, et al, eds., *Methods in Computational Physics, Volume 14: Radio Astronomy*, 1975, 108, edited

70 Elisabeth Bik, 'The Stock Photo Paper Mill,' July 5, 2020; Eva Xiao, 'Chinese Research Papers Raise Doubts, Fueling Global Questions About Scientific Integrity,' *Wall Street Journal*, July 5, 2020.

71 World Series 2016 final game sequence, ESPN.

72 Relying entirely on the words of the reported text M. NourbeSe Philip, *Zong!*, Wesleyan University Press, 2008

72 it / is said / has been decided M. NourbeSe Philip, #12 and #20 in *Zong!*, 2008, 21, 35; Wesleyan University Press, Middletown, CT; legal decision *Gregson v Gilbert* (1783) in Henry Roscoe, eds., *Reports of Cases Argued and Determined in the Court of King's Bench*, volume 3, 1831, 232 - 235

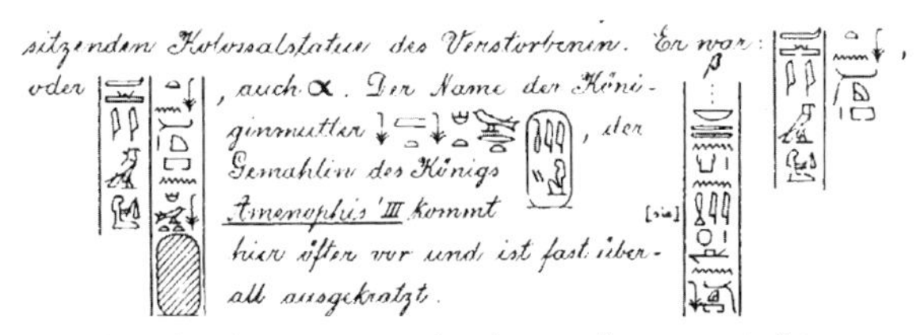

73 Karl Richard Lepsius, *Denkmäler aus Ägyptenund Äthiopien*, 1895, 138, 259

74 Earnest Alfred Wallis Budge, hieroglyphic text, Middle Egyptian 'Book of the Dead of Hunefer,' written ~1300 BCE, *Facsimiles of the Papyri of Hunefer, Anhai, Karasher, Nechtemet*, 1899, 14, 1

74 Axel Pelster, Hagan Kleinert, Michael Bachman, *Functional Closure of Schwinger-Dyson Equations in Quantum Electrodynamics*, arXiv.org, 2001, 27

75 Katsushika Hokusai, *One Hundred Views of Mt. Fuji*, 1834, 2v-3r; Smithsonian Libraries

75 Beethoven, *Grosse Fugue for piano, 4 hands*, autograph manuscript, 1826, Juilliard Manuscript Collection. Alex Ross, 'Great Fugue: Secrets of a Beethoven Manuscript' *The New Yorker*, February 6, 2006

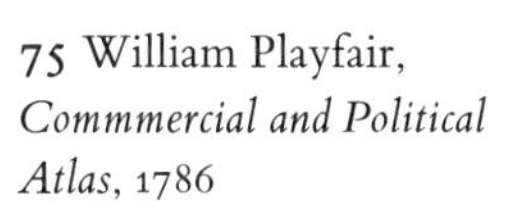

75 William Playfair, *Commmercial and Political Atlas*, 1786

Thomas preserved the main points, did not change the text. Done very well indeed! He made it easier for others.
Your kind majesty, receive this small gift, which is like a little friend.
manuscript abstract

78 Thomas Le Myésier, *Breuiculum ex Artibus Raimundi electum*, France, 1325; Karlsruhe, Badische Landesbibliothek, edited/translated

78 Mathurin Jaques Brisson, 'Cockatoo,' *Ornithologie ou méthode contenant la division des oiseaux en ordres, sections, genres, espèces & leurs variétés*, 1760, Tom. IV, Pl. XXI

Now
Bennie
has no
money (NONE) to buy Buster
a new ball which, you will remember, Pete
ate
many
letters
ago.

79 Maira Kalman, *What Pete Ate from A to Z*, 2001

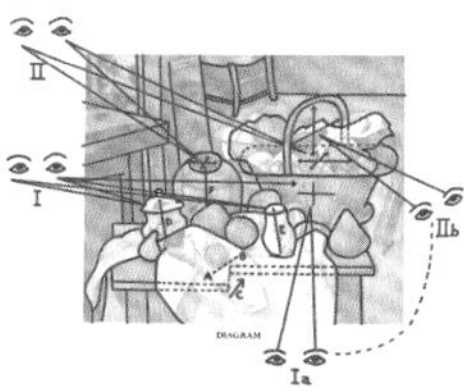

79 The metaphorical model of Cubism is the diagram John Berger, *The Moment of Cubism*, 1969, 20. Erle Loran, *Cézanne's Composition: Analysis of His Form, with Diagrams and Photos of his Motifs*, 1943

4 DATA ANALYSIS WHEN THE TRUTH MATTERS: ON THE RELATIONSHIP BETWEEN EVIDENCE AND CONCLUSIONS REMODELING STATISTICAL PRACTICE/TEACHING

80 The first principle is that you must not fool yourself
Richard Feynman, 'Cargo Cult Science,' *Surely You're Joking, Mr. Feynman!*, 1985, 343

80 Living organisms are historical structures
François Jacob, 'Evolution and Tinkering,' *Science* 196, 1977, 1166

80 Replicative mutations associated with stem cell divisions
Cristian Tomasetti, Lu Li, Bert Vogelstein, 'Stem cell divisions, somatic mutations, cancer etiology, cancer prevention,' *Science* 335, 2017. Cristian Tomasetti, Bert Vogelstein,'On the regression slope between stem cell divisions and cancer risk, and the lack of correlation between stem cell divisions and environmental factors-associated cancer risk,' *PLOS ONE* 12, 2017, 1, edited

80 When we say that Napoleon commanded armies
Leo Tolstoy, *War and Peace*, 1869, Epilogue II.VI, 1970

81 A mistake in the operating room Andrew Vickers, 'Interpreting Data From Randomized Trials,' *Nature Clinical Practice Urology* 2.9, 2005, 404–405

81 It is a principle that shines impartially on the just and the unjust
Van Wyck Brooks, *America's Coming-of-Age*, 1915, 20

81 Bias can creep into the scientific enterprise in all sorts of ways
Colin Begg, Twitter, 17 January 2019

81 Marcia Angell, 'Drug Companies & Doctors: A Story of Corruption,' *New York Review of Books*, January 19, 2009

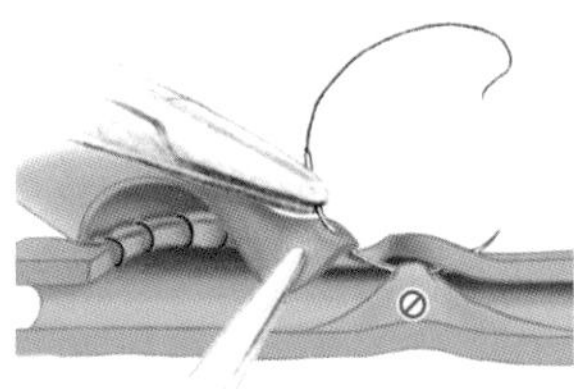

82 'Inadvertent Suturing of Posterior Wall,' Siavosh Khonsari and Colleen Sintek, *Cardiac Surgery: Safeguards and Pitfalls in Operative Technique, Third Edition*, 2003, 157, edited

83 Precise fulfillment of randomization protocols John B. Carlisle, 'Data fabrication and other reasons for non-random sampling in 5087 randomised, controlled trials in anaesthetic and general medical journals,' *Anaesthesia* 72, 2017, 931-935; Arnav Agarwal and John Ioannidis, 'PREDIMED trial of Mediterranean diet: retracted, republished, still trusted?' *BMJ* 364, 2019

83 Roy Lichtenstein, *Image Duplicator*, 1963, oil and magna on canvas

83 Images from 20,621 papers published Elisabeth M. Bik, Arturo Casadevall, Ferric C. Fang, 'The Prevalence of Inappropriate Image Duplication in Biomedical Research Publications,' *American Society for Microbiology mBio* 7, 2016, 809-816

83 We analyzed 960 papers published Elisabeth M. Bik et al, 'Analysis and Correction of Inappropriate Image Duplication: The *Molecular and Cellular Biology* Experience,' *American Society for Microbiology mBio* 10, 2018, 309-318

83 A programmatic scan of leading genomics journals Mark Ziemann, Yotam Eren, Assam El-Osta, 'Gene name errors are widespread in the scientific literature,' *Genome Biology* 17, 2016

84 Yves Klein, photograph Harry Shunk and János (Jean) Kender, 'Leap into the void,' 1960, Metropolitan Museum of Art; Mia Fineman, *Faking It, Manipulated Photography Before Photoshop*, 2012, 182-183

85 Can the data presented be traced back to primary data 'Report of the investigation committee on the possibility of scientific misconduct in work of Hendrik Schön and coauthors,' Appendix E: Elaborated final list of allegations, Lucent Technologies, September 2002, E3-E4

86 Bang-bang duplicates Cuthbert Daniel, in Edward Tufte, 'A Conversation with Cuthbert Daniel,' *Statistical Science* 3, 1988, 419; Bahar Gholipour, 'Statistical errors may taint as many as half of mouse studies,' *Spectrum*, 15 March 2018

86 Randall Munroe, 'All Sports Commentary,' XKCD, 27 May 2011

86 Ignorance more frequently begets confidence Charles Darwin, 'Introduction,' *The Descent of Man*, 1871, 3

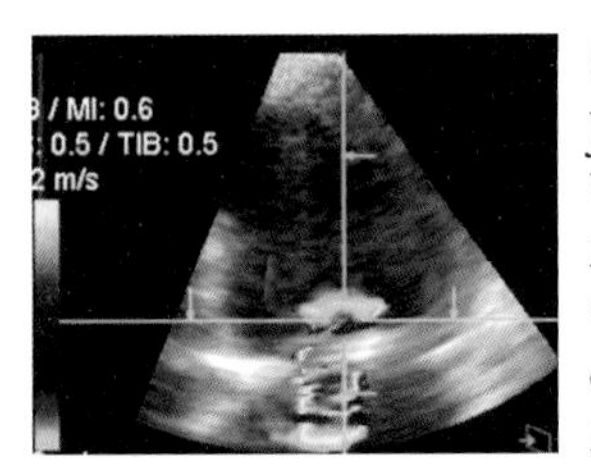

88 Massimiliano Meineri, Wendy Tsang, Jodi Crossingham, TEE, Philips EPIQ 7 Ultrasound System, *Virtual Transesophageal Echocardiography*. Alexandra Gonçalves, Carla Sousa, José Zamorano, 'Assessment of The Mitral Valve in the 3D Era,' *British Society of Echocardigraphy*, 2018

89 Acoustic Artifacts in 2D Imaging Acquisition TEE technical manual, Philips EPIQ 7 Ultrasound System User Manual, 3.0, August 2013, Koninklijke Philips NV, 179-183

90 It is the experience of statisticians Frederick Mosteller, 'A Pathbreaking Report' in *On Equality of Educational Opportunity*, Frederick Mosteller and Daniel P. Moynihan, 1972, 33-34, edited

90 Far better an approximate answer to the right question John Tukey, 'The Future of Data Analysis,' *AMS* 33, 1962, 13

90 Simple methods typically yield performance almost as good David J. Hand, 'Classifier Technology and the Illusion of Progress,' *Statistical Science* 21, 2006, 1

91 Values are missing Chris Groskopf, 'The Quartz Guide to Bad Data,' https://github.com/Quartz/bad-data-guide

91 Of all the technical debt you can incur Jeff Atwood, Twitter, 10 April 2019, edited **Going through some old data/code** Nate Silver, Twitter, 18 July 2018, edited. **Bioinformatics . . . or 'advanced file copying'** Nick Loman, Twitter, 29 January 2014

92 Less healthy patients may never arrive for treatment Darrel P. Francis, Twitter, 4 March 2018

94 The fallacy of equivocation occurs David Hackett Fischer, *Historians' Fallacies*, 1970, 274, 'Fallacies of semantical distortion'

94 Most medieval castles were made of wood John D. Cook

⚠ IMMORTAL TIME BIAS ⚠ SKETCHY IN, SKETCHY OUT ⚠ BIAS IN, BIAS OUT

98 A journey of a thousand hypotheses begins with a single stepwise regression Matthew Hankins, Twitter, 23 October 2017

99 Lack of independence within a sample 'Does your data violate Kaplan-Meier assumptions?' JnF Specialities, LLC, Colorado Springs, edited

group 1
group 2
time

Range from lowest to highest amount
Median
Interquartile Range

100 Mary Eleanor Spear, *Charting Statistics*, 1952, 164-6

100 Boxplot variations Justin Matejka and George Fitzmaurice, 'Same Stats, Different Graphs,' Autodesk Research, Toronto Ontario Canada, May 2017, edited; Robert McGill, John W. Tukey, Wayne A. Larsen, 'Variations of Box Plots,' *The American Statistician* 32, 1978, 12-16, J.W. Tukey, *Exploratory Data Analysis*, 1970, 1977, 39-41; ET, *The Visual Display of Quantitative Information*, 1983, 2001, 123-125

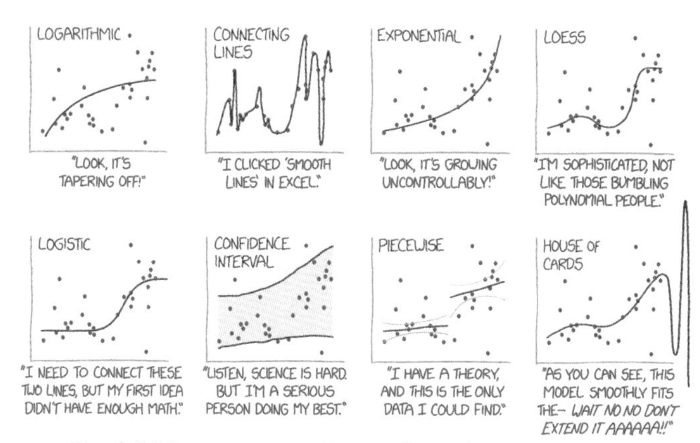

LOGARITHMIC
"LOOK, IT'S TAPERING OFF!"
CONNECTING LINES
"I CLICKED 'SMOOTH LINES' IN EXCEL."
EXPONENTIAL
"LOOK, IT'S GROWING UNCONTROLLABLY!"
LOESS
"I'M SOPHISTICATED, NOT LIKE THOSE BUMBLING POLYNOMIAL PEOPLE."
LOGISTIC
"I NEED TO CONNECT THESE TWO LINES, BUT MY FIRST IDEA DIDN'T HAVE ENOUGH MATH."
CONFIDENCE INTERVAL
"LISTEN, SCIENCE IS HARD. BUT I'M A SERIOUS PERSON DOING MY BEST."
PIECEWISE
"I HAVE A THEORY, AND THIS IS THE ONLY DATA I COULD FIND."
HOUSE OF CARDS
"AS YOU CAN SEE, THIS MODEL SMOOTHLY FITS THE- WAIT NO NO DON'T EXTEND IT AAAAAA!!"

105 Randall Munroe, XKCD, 'Curve-fitting,' September 19, 2018, xkcd.com/2048

106 No matter how big one's proton detector
Natalie Wolchover, 'Grand Unification Dream Kept at Bay,' *Quanta Magazine*, 15 December 2016

106 An unbelievably large literature William Feller, *An Introduction to Probability Theory and Its Applications* II, 1971, 52 - 53

106 If someone shows you simulations that only show the superiority
Byran Smucker and Rob Tibshirani, Twitter, July 2018

106 Tuning your own method but insufficiently tuning other methods
Manjari Narayan and Rob Tibshirani, Twitter, 29 September 2018

106 Wizard of Oz, Mirko Ilic Corp., for Dave Kehr, 'When a Cyberstar Is Born,' *The New York Times*, 18 November 2001

106 In that Empire, the Art of Cartography Jorge Luis Borges, 'On Exactitude in Science,' *A Universal History of Infamy* (*Historia universal de la infamia*), 1946; translated by Andrew Hurley, *Collected Fictions*, 1960

107 The clinical trial compared PTCA John Mandrola, 'Think you are immune from bias? Think again,' *Medscape*, 16 November 2018

109 On the difference between machine learning and AI
Mat Velloso, Twitter, 22 November 2018

110 Thomas Chalmers, a founder of evidence-based medicine
N.D. Grace, H. Muench, T.C. Chalmers, 'The Present Status of Shunts for Portal Hypertension in Cirrhosis,' *Gastroenterology* 50, 1966, 684-691; ET, *Beautiful Evidence*, 2006, 145, revised

110 Why Most Published Research Findings Are False,
John P.A. Ioannidis, *Public Library of Science Medicine* 2.8, 2005

111 Meta-researchers do relevant and stunning replication studies
C. Glenn Begley and Lee M. Ellis, 'Raise standards for preclinical cancer research,' *Nature* 483, 2012, 531 - 533; 46 co-authors and Brian Nosek, 'Many analysts, one dataset: Making transparent how variations in analytical choices affect results,' Zurich Open Repository and Archive, 2017; Colin Camerer, Brian Nosek, et al, 'Evaluating the replicability of social science experiments in *Nature* and *Science* between 2010 and 2015,' *Nature Human Behavior* 2, 2018, 637 - 644

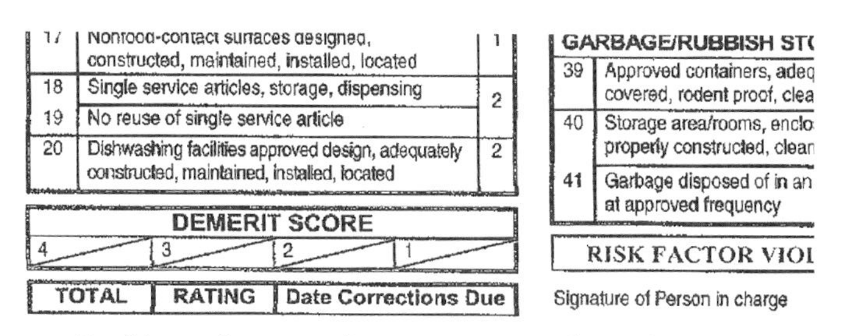

17	Nonfood-contact surfaces designed, constructed, maintained, installed, located	1
18	Single service articles, storage, dispensing	2
19	No reuse of single service article	
20	Dishwashing facilities approved design, adequately constructed, maintained, installed, located	2

DEMERIT SCORE			
4	3	2	1

TOTAL	RATING	Date Corrections Due

GARBAGE/RUBBISH ST(	
39	Approved containers, adeq covered, rodent proof, clea
40	Storage area/rooms, enclo properly constructed, clear
41	Garbage disposed of in an at approved frequency

RISK FACTOR VIOI

Signature of Person in charge

112 Food inspection report form Department of Health, State of Connecticut, EHS-106, Regulations of Connecticut State Agencies, Title 19, Public Health and Safety, Chapter II, revised May 2018, 69

114 People of the same trade seldom meet together
Adam Smith, *The Wealth of Nations*, 1776, book I, chapter X

114 It is difficult to get people to understand Upton Sinclair, *I, Candidate for Governor: And How I Got Licked*, 1934, 109, edited

114 I've worked with 1000s of experts in psych diagnosis
Alan Cassels, Twitter, 11 February 2020

114 Marcia Angell, 'Drug Companies and Doctors: A Story of Corruption,' *New York Review of Books*, January 19, 2009

114 Guidelines from professional societies are increasingly influential
John P.A. Ioannidis, 'Professional Societies Should Abstain From Authorship of Guidelines and Disease Definition Statements,' *Circulation: Cardiovascular Quality and Outcomes* 11, 2018, 109, 1 - 4

116 One day when I was a junior medical student
E.E. Peacock, Jr., University of Arizona College of Medicine; quoted in *Medical World News*, 1 September 1972, 45

116 I'm telling you, as someone who works with patients
Dr. Lisa DeAngelis, quoted in Katie Thomas and Charles Ornstein, 'Memorial Sloan Kettering's Season of Turmoil,' *The New York Times* and *ProPublica*, 31 December 2018

116 Bias can creep into the scientific enterprise in all sorts of ways
Colin Begg, Twitter, 17 January 2019

116 The key substantive issue is that the problems we face
Colin Begg, quoted in Katie Thomas and Charles Ornstein, 'Memorial Sloan Kettering's Season of Turmoil,' *The New York Times* and *ProPublica*, 31 December 2018

116 The most important medical advance in our generation Vinayak Prasad, *Malignant: How Bad Policy and Bad Evidence Harm People with Cancer*, 2020

117 Epidemiologic Signatures in Cancer, Gilbert Welch, Barnett S. Kramer, William C. Black, *NEJM*, Oct 2019

5 ANNOTATIONS: EXPLANATORY WORDS, NUMBERS, GRAPHICS, IMAGES PLACED ON LOCAL GRIDS

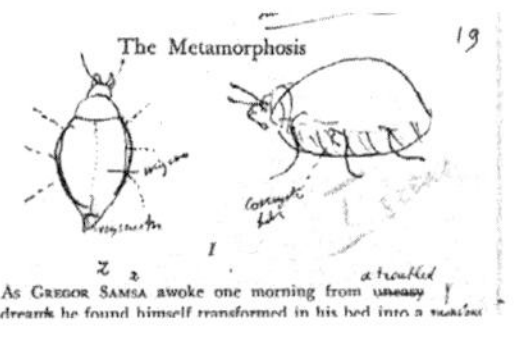

120 Franz Kafka, *The Metamorphosis* (*Die Verwandlung*), 1915, translation by A.L. Lloyd, 1946; Vladimir Nabokov's annotated copy, New York Public Library, Berg Collection

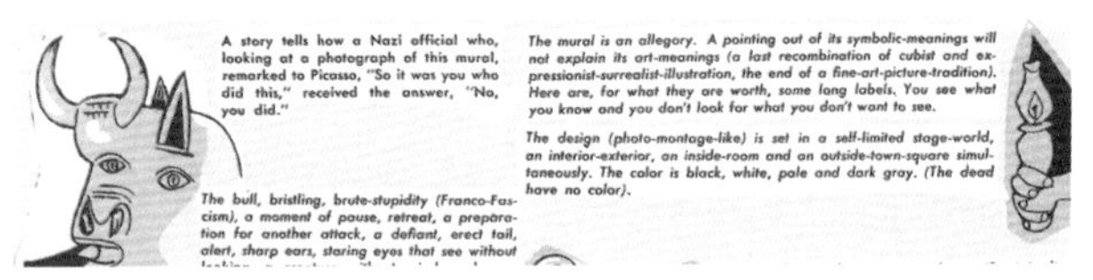

121 Athanasius Kircher, *Musurgia universalis*, 1650; St. Andrews University Special Collections, f. 30

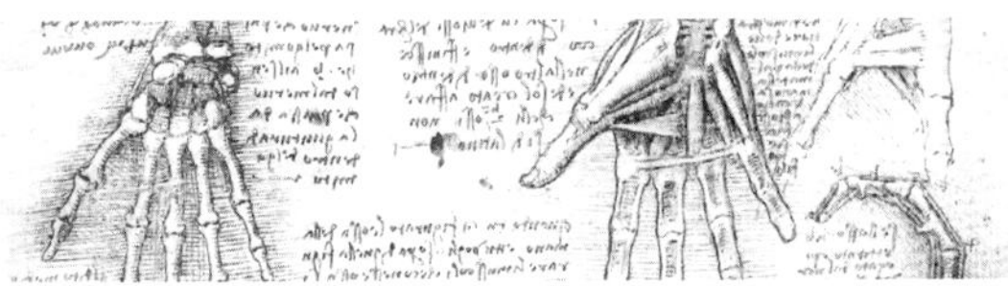

122-123 Ad Reinhardt, 'How to Look at a Mural,' 1947, offset lithograph, 194 × 272 mm, Univ of Kansas, Spencer Museum of Art; Pablo Picasso, *Guernica*, Madrid, Museo Reina Sofia

123 No, painting is not done to decorate Pablo Picasso, quoted in Alfred Barr, *Picasso: Fifty years of his art*, 1946, 250

126 Leonardo da Vinci, upper extremity, *Bones and Muscles of the Arm*, 1510 - 1511, pen and ink with wash, 29.3 × 20.1 cm, Windsor Royal Library, sheet no. 19000v, edited; *The Drawings of Leonardo da Vinci in the Collection of Her Majesty the Queen*, edited by Kenneth Clark, Carlo Pedretti, 1968; translation adapted *Leonardo on the Human Body*, translated edited by Charles D. O'Malley, J.B. de C.M. Saunders, 1983

127 Megan Jaegerman 1990 - 1998 portfolio, *The New York Times* adapted from *The Wellness Letter*. Edited, reconstructed.

128 *Mugiz al-Qanun* (*Epitome of the Canon of Medicine of Avicenna*), attributed to Ala-al-din abu Al-Hassan Ali ibn Abi-Hazm al-Qarshi al-Dimashqi (Ibn al-Nafis), 1213 - 1288; Persian translation of original Arabic text, marginal annotations in Persian and Arabic; London, Wellcome Library, MS Arabic 199, ff. 1r, 45r, 69r, 103r, identified with assistance from Nahyan Fancy

128 Opening page of Tractate Berakhot: Hebrew text from *Koren Talmud Bavli, vol.1: Berakhot, Hebrew/English*, with commentary by Rabbi Adin Steinsaltz, edited and adapted by Rabbi Shalom Berger, 2012.

6 INSTRUCTIONS AT POINT OF NEED

130 Envelope addressed to 'Danish woman who works in a supermarket in Búðardalur,' a remote coastal village in northwest Iceland, mailed by a tourist in Reykjavík unable to find exact address; 'Frumleg áritun sendibréfs en dugði þó,' *Skessuhornið* 20 May 2016, photo Steina Matt

130 E.G. Lutz, *What to Draw and How to Draw It*, 1913, 11, edited

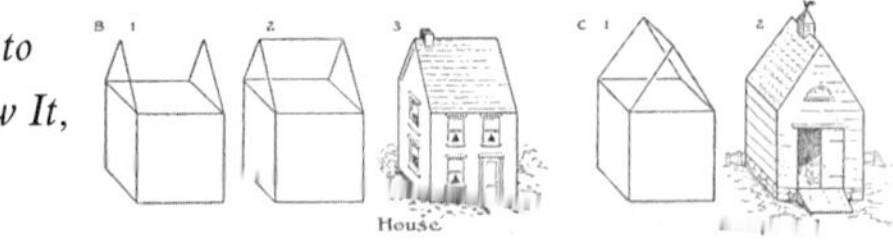

130 ET, *Avoid Not Writing* and *Avoid Not Coding*, digital prints on canvas, 2013

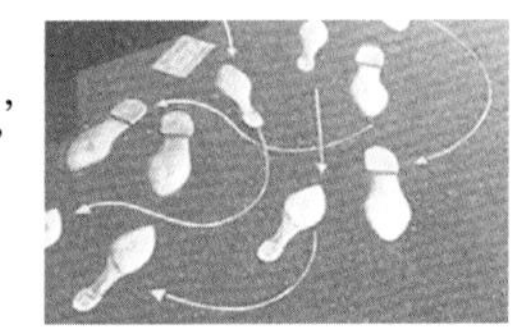

131 Jack Mackie, 'Dance Steps on Broadway,' 1982, embedded in sidewalk on Broadway Street, Seattle; photo Erik Schwab, 2005

131 Nagasaki Prefectural Art Museum, signage plan, designed by Kenya Hara and Yoshiaki Irobe, Hara Design Institute, 2005

一丨丶丿乙亅人八力匸十厶又口子巛辵彡彳

131 Chinese character stroke sequences, Wikipedia Commons Stroke Order Project; based on traditional Chinese, 1995, produced by the Taiwan Ministry of Education, 1997, edited

132 'Fish Smart,' poster designed by Lauren Wohl-Sanchez for the California Department of Public Health, photo ET at Fort Mason on San Francisco Bay, 23 December 2017

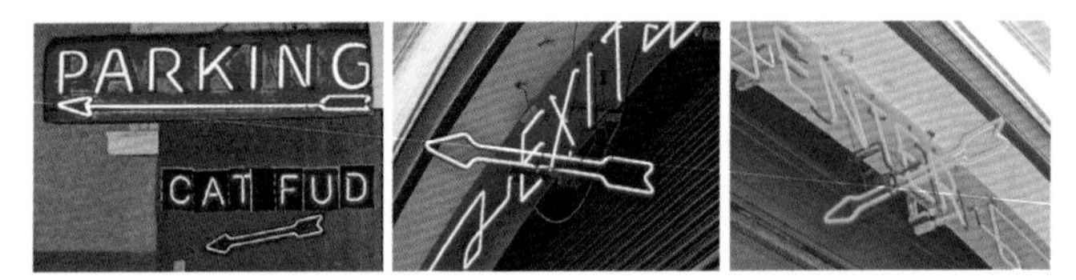

132 San Francisco Palace Hotel Garage, ET photos (cat fud photoshopped in)

133 Manuscript copied by a scribe named Mauritius at monastery of Mont St Michel, ~1075-1100 CE; text Ado of Vienne, *Chronicon de sex aetatibus mundi*, edited in Migne, *Patrologia Latina*

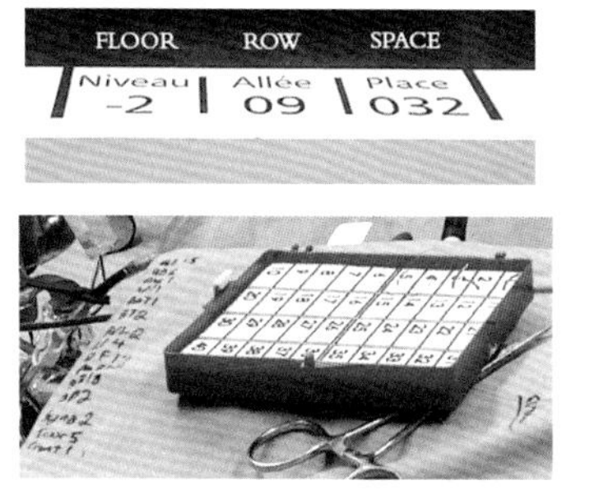

133 Parking space location label, Paris Charles de Gaulle Airport, photo by Tchad

133 Mitral valve repair, inventory tracking of surgical instruments, Cleveland Clinic, May 2017

134-135 Joseph Heller, *Catch 22*, handwritten plotchart, 1954, edited

134 Laurence Sterne, detail from original printing of *The Life and Opinions of Tristram Shandy*, volume VI, 1762; London, British Library Ashley 1770, 152 - 153

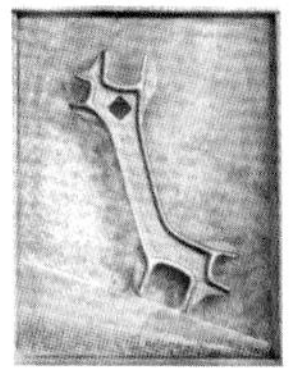

136 ET, *Art is Art*, diamond sign, 201

136-137 See Now . . . Words Later ET, *Seeing Around*, 2010, 30-31, edited. ET, *Open-Ended Wrench*, 2008, steel

137 Do you see yonder cloud that's almost in shape of a camel? William Shakespeare, *Hamlet*, Act III, Scene II, lines 339-344

137 Rage for wanting to conclude Gustave Flaubert, letter to Mlle Leroyer de Chantepie, 23 October 1865, Flaubert, *Correspondance*, ed. Louis Conard, 1929, volume V, 111, translated by Dawn Finley

137 ET, philosophical diamond signs, Hogpen Hill Farms, 2011-2019

138 Asaf Degani, 'On the Typography of Flight-Deck Documentation.' NASA, December 1992

7 LISTS: THEORY AND PRACTICE

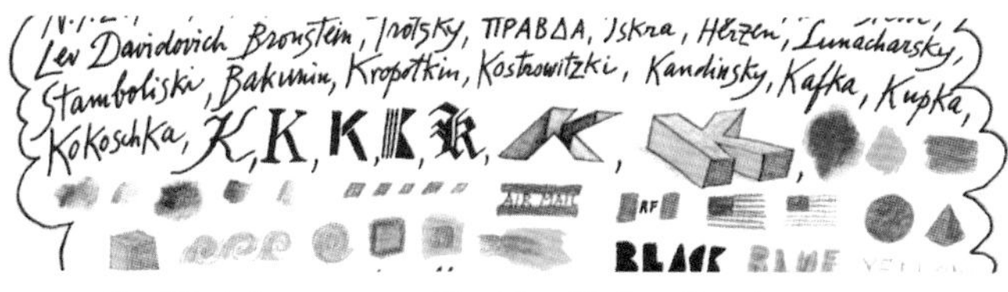

140 Saul Steinberg, cover, *The New Yorker*, October 18, 1969

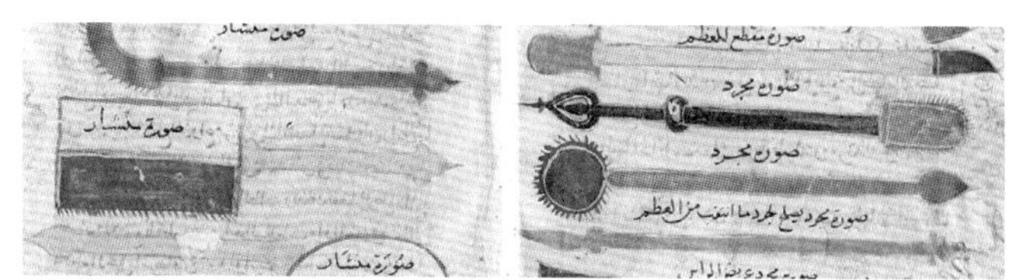

141 Abu al-Qasim Khalaf ibn al-Abbas Al-Zahrawi (Abulcasis), *Kitab al-tasrif*, 'Method of Medicine,' chapter 30 on surgery, Córdoba, Spain ~1000 CE; copy on paper, 12th century, Leiden ms. Or. 2540

142 Those that belong to the emperor Jorge Luis Borges, 'Celestial Emporium of Benevolent Knowledge' ('*Emporio celestial de conocimientos benévolos*'), 1942; *Borges: Selected Nonfictions*, ed. Eliot Weinberger, translated by Esther Allen and Suzanne Jill Levine, 2000

143 A long, long time ago Martin Hardee, 'Tufte Story: AnswerBook,' blogs.sun.com, 24 June 2004

143 Aelius Donatus, *Ars Maior*, manuscript, 9th-century France; Paris Bibliothèque nationale de France MSS Latin 13025, f. 11v

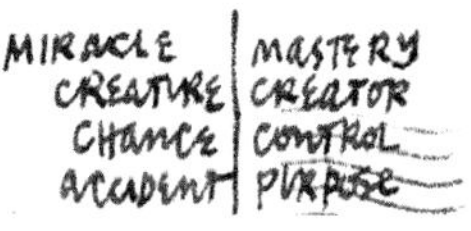

144 Ad Reinhardt, 'Variety of Dualisms,' postcard to Katherine Scrivener, 16 April 1951

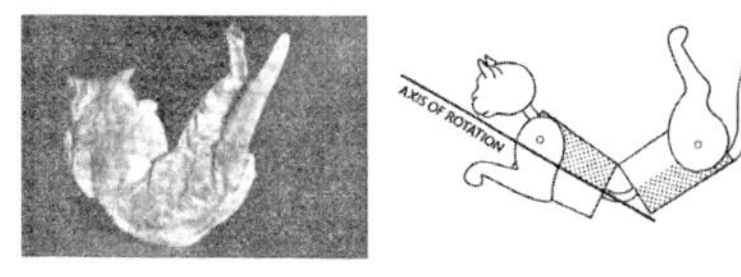

144 Donald McDonald, 'How does a cat fall on its feet?' *The New Scientist* 7, 1960, 1647-1649

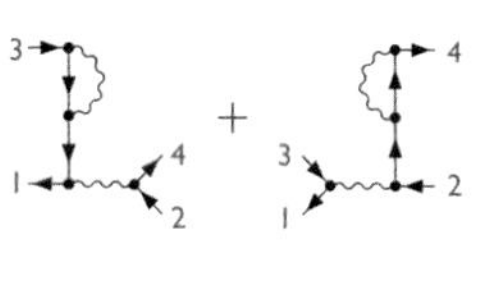

145 Axel Pelster, Hagan Kleinert, Michael Bachman, *Functional Closure of Schwinger-Dyson Equations in Quantum Electrodynamics*, arXiv.org, 2001, 27

149 F. Scott Fitzgerald emphasizes the miscellaneousness of the list David Lodge, commentary on Fitzgerald's lists in *The Art of Fiction: Illustrated from Classic and Modern Texts*, 1992, 64 - 65

8 REMODELING PRESENTATIONS, MEETINGS, TEACHING, BOOK PUBLISHING

150 Annotated map of Paris, adapted from Michel Etienne Turgot and Louis Bretez, *Plan de Paris* (1739), plate II

151 It's not what you say, it's what they hear Red Auerbach, former coach of the Boston Celtics basketball team, NPR BROADCAST 2004

151 The single biggest problem in communication is the illusion that it has taken place Attributed to George Bernard Shaw; also 'The great enemy of communication, we find, is the illusion of it,' William H. Whyte, 'Is Anybody Listening,' *Fortune*, 1950, 174

glucose 6.6
respiration 12
temperature 37.1°C

$64.368	Vanguard 500 Index	-2.0%	+12.2%	-11.7%	-0.8%
62.510	Fidelity Magellan	-2.1	+11.3	-12.9	-0.2
50.329	Amer A Invest Co Am	-1.2	+09.4	-03.9	+4.0
47.355	Amer A WA Mutual Inv	-1.5	+09.9	+00.8	+3.0
40.500	PIMCO Instl Tot Return	-2.3	+02.4	+09.4	+7.6
37.641	Amer A Grow Fd Amer	-2.9	+14.1	-11.0	+7.4
31.161	Fidelity Contrafund	-1.0	+10.7	-06.5	+3.0
28.296	Fidelity Growth & Inc	-1.8	+08.2	-08.7	-0.1
25.314	Amer A Inc Fund Amer	-0.5	+09.9	+05.5	+5.4
24.155	Vanguard Instl Index	-2.0	+12.3	-11.6	-0.7

153 ET, Microsoft Machine Learning and Data Science Summit

155 Steve Jobs: 'I hate the way people use slide presentations instead of thinking. People who know what they're talking about don't need PowerPoint" Walter Isaacson, *Steve Jobs*, 2015, 337

155 We have study hall at the beginning of our meetings
Jeff Bezos, television interview, Charlie Rose, 16 November 2012

9 A VISUAL INDEX OF SOURCES AND IMAGES REMODELS THE BACK MATTER IN BOOK

(a recursive quilt-quilt visual index for chapter 9)

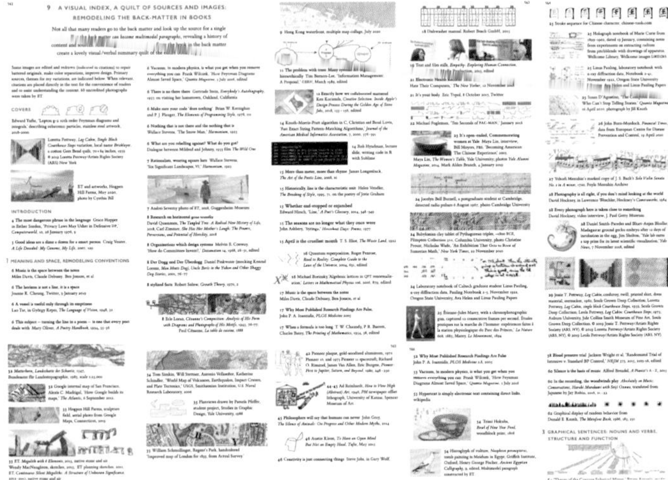

The following artworks, illustrations, photographs, quotations are reprinted by permission. All rights are reserved by the claimed copyright holders listed below, with artist reproduction rights reserved by VAGA, SPADEM, ARS, ADAGP, where relevant. Permission to reproduce any of these artworks, illustrations, photographs, quotations must be obtained directly from copyright holders. Every effort has been made to trace all copyright owners. Special extended efforts were required by the pandemic. Graphics Press apologizes for any omissions and will remedy such cases in any future editions.

BOOK COVERS © 2019 Loretta Pettway/Artists Rights Society (ARS), New York Feynman diagram and sculpture photographs © Edward Tufte

INTRODUCTION © J. Craig Venter, 2007. All rights reserved

CHAPTER 1 MEANING AND SPACE © Joanne K. Cheung, 2019. © Frank Wilczek, *How Feynman Diagrams Almost Saved Space*, 2016. *Cezanne's Composition: Analysis of His Form with Diagrams and Photographs of His Motifs*, Erle Loran, © 1980 by the Regents of the University of California. Published by the University of California Press. Paul Cézanne, *Still life with basket, or the Kitchen Table* © Musée d'Orsay, Paris/Réunion des Musées Nationaux/Art Resource, New York. Tim Berners-Lee internet proposal 1989 CERN © 2008. © Ken Kocienda, *Creative Selection: Inside Apple's Design Process During the Golden Age of Steve Jobs*, 2018. Illustration by Guy Shield © 2018-2019 Ken Kocienda. *The Breaking of Style: Hopkins, Heaney, Graham* by Helen Vendler, Cambridge, Mass.: Harvard University Press, © 1995 by the President and Fellows of Harvard College. Four lines from "Syringa" from Houseboat Days by John Ashbery. Copyright © 1975, 1976, 1977, 1999 by John Ashbery. Reprinted by permission of Georges Borchardt, Inc., on behalf of the author. © 2004 by Roger Penrose. Reprinted by permission from Springer Nature Customer Service Center GmbH: Springer *Nature: Letters in Mathematical Physics*, Algebraic Lattices in QFT Renormalization, Michael Borinsky, © 2016. Cleveland Clinic Media Production © 2013. © 2006, Edward Tufte, animals scatterplot, *Beautiful Evidence*. © 2018 Michael Fogleman, Ten Seconds of PAC-MAN. *The Women's Table* © 2018 Yale News Maya Lin table © 2014 Yale Alumni Magazine Pulsar Chart, Churchill Archives Centre, HWSH Acc 355 Babylonian clay tablet © Columbia University Special Collections. © 2010 Photo Christine Proust, Nicholas Wade. Ava Helen and Linus Pauling Papers © Oregon State University Libraries Special Collections, Archives Research Center. Wellcome Library, London, Wellcome Images, wellcomeimages.org, Marie Curie notebook 27 May 1899-4 December 1902. containing notes of experiments, etc. on radioactive substances copyrighted work under Creative Commons Attribution only license CC BY 4.0 © 2020 photo by Jill Knuth John Burn-Murdoch, 2020 "Daily death tolls are still accelerating in many countries," The Financial Times/FT.com, May 2, 2020. Used under license The Financial Times. All rights reserved. Swiss maps, Source: Federal Office of Topography swisstopo, © swisstopo. Menuhin's copy of Sonata © 168490-1001 from Foyle Menuhin Archive, Royal Academy of Music. David Hockney "Pearblossom Hwy., 11-18th April 1986" © David Hockney Collection J. Paul Getty Museum, Los Angeles. Gecko embryo © Daniel Smith-Paredes, Bhart Anjan-Bhullar, *Yale News* 2018. © 2019 Estate of Leola Pettway/Artists Rights Society (ARS), New York, © 2019 Jessie T. Pettway/Artists Rights Society (ARS), New York. © Jeffrey T. Leek, "Batch effects in high-throughput data," 2010. © 2012 "How Google Builds Its Maps," *The Atlantic*, Alexis C. Madrigal. © 2006 USGS, Smithsonian Institution, U.S. Naval Research Laboratory. © The Morgan Library & Museum. MA 6305.4.10. Gift of Laurent de Brunhoff; 2004. Photographic credit: The Morgan Library & Museum, New York © 2016 Fényes Lóránd, *Big Dipper, Deep Sky*. © 2015 Martin Vargic, Halcyon Maps. © Saul Steinberg Foundation/Artists Rights Society (ARS) New York *How to View High (Abstract) Art* © 2020 Estate of Ad Reinhardt/Artists Rights Society (ARS), New York. © 2013 by John Gray, all rights reserved. © 2015 by Austin Kleon, all rights reserved. *The Principles of Uncertainty* by Maira Kalman © 2007 Maira Kalman. By permission of Penguin Press, an imprint of Penguin. Publishing Group, division of Penguin Random House LLC, all rights reserved.

CHAPTER 2 RESPONSIVE TYPOGRAPHY W Somerset Maugham, "The Book Bag." © 1933 United Agents, Ltd. Printed with permission. © John Ioannidis (2005) "Why most published research findings are false," *PLoS Med.* "Il Pleut." *Calligrammes: Poems of Peace and War.* Reprinted with permission of University of California Press, © 2004. © Metropolitan Museum of Art, Teisai Hokuba, *Bowl of New Year Food.* H.O. Havemeyer Collection, Bequest of Mrs. H.O. Havemeyer, 1929. Khonsari, Siavosh and Sintek, Colleen, MD; *Cardiac Surgery: Safeguards and Pitfalls in Operative Technique*, Third Edition; Wolters Kluwer Health. Lippincott Williams & Wilkins © 2003 © Association for Vascular Access, Dr. Peter Lax, 2019 © 2014 by Roche Diacnostics GmbH. All rights reserved. Excerpt *The War Against Cliché: Essays and Reviews 1971-2000* by Martin Amis. Copyright © 2001 by Martin Amis, used by permission of The Wylie Agency L.L.C. Excerpt from *Absolutely on Music; Conversations* by Haruki Murakami with Seiji Ozawa and translated by Jay Rubin, copyright © 2016 Seiji Ozawa and Haruki Murakami reprinted by permission of AnchorCananda/ Doubleday Canada, a division of Penguin Random house Canada Limited. All rights reserved. Any third party use of this material, outside of this publication is prohibited. Interested parties must apply directly to Penguin Random House Canada Limited for permission. Ancient Egyptian Calligraphy. © 1979, 1983, 1988, 1999 by Henry George Fischer. All rights reserved. Printed with permission. *The Work, Computer Age Statistical Inference*, first published by Cambridge University Press, © Bradley Efron and Trevor Hastie 2016. All rights reserved. FT Graphics: John Burn-Murdoch, 2020 "Cumulative number of deaths, by number of days since the 10th death," The Financial Times/FT.com, 22 March. Used under license from *The Financial Times*. All rights reserved.

CHAPTER 3 GRAPHICAL SENTENCES © 2002 Universiteit Gent, *Galileo, the Telescope, and the Science of Optics in the XVI.* © 1957 Disney *The Universe and Forever* © 2003. *Here with you*, print © Wendy MacNaughton. Methods in Computational Physics: Advances in Research and Applications, 14, Timothy H. Hankins and Barney J Rickett, *Pulsar Signal Processing*, 55-129, Copyright © 1975, with permission from Elsevier Books. "Clustering of galaxies in the Baryon Oscillation Spectroscopic Survey" © 2014 Lauren Anderson, et al. Reprinted Table II, "Constraining a Thin Dark Matter Disk with Gaia" permission from Shutz, Katelin, et al, *Physical Review Letters;* vol. 121, 081101 (2018), © 2018 by the American Physical Society. © *Zong!* Wesleyan University Press, Middletown, CT. All rights reserved. © 2008 by M. NourbeSe Philip, *Zong!* All rights reserved. Birtwistle © Copyright 1972 by Universal Edition (London) Ltd., London/UE 15331 All rights reserved © Elaine Gould, 2011. Reprinted from Annals of Physics, Vol 297/2, Axel Pelster, Hagen Kleinert, Michael Bachmann, "Functional Closure of Schwinger-Dyson Equations in Quantum Electrodynamics 1. Generations of Connected and One-Particle Irreducible Feynman Diagrams," pp. 363-395, copyright © 2002, with permission from Elsevier Image © The Metropolitan Museum of Art, Katsushika Hokusai, *One Hundred Views of Mount Fuji.* Ludwig van Beethoven, *Grosse Fuge in B flat major.* Courtesy of the Julliard Manuscript Collection Copyright Republished with permission of Princeton University Press, from *The Hedgehog and the Fox: An essay on Tolstoy's view of history*; Isaiah Berlin, Isaiah and Henry Hardy, 2013; permission conveyed through Copyright Clearance Center, Inc, revised text © Isaiah Berlin 1978. © 2015 by Maira Kalman, by permission of Penguin Press, an imprint of Penguin Publishing Group, a division of Penguin Random House LLC. All rights reserved.

CHAPTER 4 DATA ANALYSIS WHEN THE TRUTH MATTERS *The New York Review of Books* © 2009 Marcia Angell. *Image Duplicator* © 1963 Estate of Roy Lichtenstein Yves Klein photo, Getty Research Institute, Los Angeles (2014.R.20). Photograph by Shunk-Kender. © J. Paul Getty Trust. Image © The Metropolitan Museum of Art. Image Source: Art Resource, NY. Artistic action by Yves Klein © The Estate of Yves Klein c/o ADAGP, Paris 2020 Collaboration Harry Shunk and János Kender © J. Paul Getty Trust. Getty Research Institute, Los Angeles (2014.R.20) © 2011, 2018 xkcd, www.xkcd.com. Reprinted from *Journal of Clinical Epidemiology*, vol. 98; Timo B. Brakenhoff, Marian Mitroiu, Ruth H. Keogh, Karel G.M. Moons, Rolf H.H. Groenwold, Maarten van Smeden; pp 89-97 © 2018 with permission from Elsevier. From New England Journal of Medicine; H. Gilbert Welch, Barnett S. Kramer; and, William C. Black, "Epidemiologic Signatures In Cancer," 381, pp 1378-1386, © 2019 reprinted with permission from Massachusetts Medical Society. © 2017 Dr. Marc Gillnov 3D TEE © Image reproduced with permission from http://pie.med.utoronto.ca/TEE TEE technical manual © 2013 Koninklijke Philips N.V., all rights reserved. Frederick Mosteller, Daniel P. Moynihan, *On Equality of Educational Opportunity.* New York, Random House, © 1972 "The Quartz guide to bad data" © 2015 Christopher Groskopf. "Tackling the widespread and critical impact of batch effects in high-throughput data," © Jeffrey T. Leek, 2010. Progression-Free Survival: Meaningful or Simply Measurable," Christopher M. Booth, Elizabeth A. Eisenhauer, *Journal of Clinical Oncology*, 30, (April 2012) 1030-1033. Reprinted with permission. © 2012 American Society of Clinical Oncology. All rights reserved. © Rashida Richardson, Jason M. Schultz; Kate Crawford, "Dirty Data, Bad Predictions: How Civil Rights Violations Impact Police Data, Predictive Policing Systems, and Justice; vol. 94:15 *New York University Law Review*, p.15, May 2019. *"Does your data violate Kaplan-Meier assumptions?"* copyright © 2001-2020 JnF Specialties, LLC Autodesk Research © 2017 Justin Matejka and George Fitzmaurice. Reprinted with permission of AAAS from Science Advances 03 Oct 2018: Vol. 4, no. 10, eaav1784 (DOI: 10.1126/sciadv, aav1784) © Alex Teachey and David M. Kipping, some rights reserved; exclusive licensee American Assoc. for the Advancement of Science. Distributed under a Creative Commons Attribution Non-Commercial License 4.0 (CC BY-NC). Brian A. Nosek, R. Silberzahn, E.L. Uhlmann, D.P. Martin, et al, *Advances in Methods and Practices in Psychological Science*, vol. 1(3), pp 337-356, copyright © 2018 by The Authors, reprinted by permission of SAGE Publications, Inc. © 2016 Natalie Wolchover, *Quanta Magazine. Wizard of Oz* © Mirko Ilic Corp. *Wizard of Oz* © Everett Collection, Inc. "On Exactitude in Science," © 1998 by Maria Kodama; translation © 1998 by Penguin Random House LLC, from Collected Fictions: vol. 3 by Jorge Luis Borges, translated by Andrew Hurley, used by permission of Viking Books, an imprint of Penguin Publishing Group, a division of Penguin Random House LLC. All rights reserved. "Think You're Immune From Bias?"© John Mandrola, *Medscape* 2018. "Professional Societies Should Abstain rom Authorship of Guidelines and Disease Definition Statements" © 2018 John P.A. Ioannidis

CHAPTER 5 ANNOTATIONS Annotated *The Metamorphosis* Copyright © Vladimir Nabokov, used by permission of The Wylie Agency LLC. *Musurgia universalis*, Athanasius Kircher, © St. Andrews University Special Collections. *How to Look at a Mural*, Ad Reinhardt © 2020 Estate of Ad Reinhardt/Artists Rights Society (ARS), New York. Annotated *Pride and Prejudice* © 2004 by David M. Shapard. Leonardo da Vinci upper extremity, Royal Collection Trust/© Her Majesty Queen Elizabeth II 2020 Translation *Leonardo on the Human Body* © 1983 Charles D. O'Malley and J.B. de C.M. Saunders. Interlinear Persian translation, London, Wellcome Library, ms Arabic 199. Talmud annotation © 1984 by Barry W. Holtz. All rights reserved.

CHAPTER 6 INSTRUCTIONS AT POINT OF NEED Photo, Iceland envelope © 2016 Steina Matt Jack Mackie © Dance Steps on Broadway 1982, cast bronze inlay, Seattle Arts Commission. Photo credit: Art on File. Nagasaki Prefectural Art Museum © 2005 Kenya Hara and Yoshiaki Irobe, Hara Design Institute. *Chronicon de sex aetatibus mundi* © Ado of Vienne *Catch 22* plotchart, Robert D. Farber University Archives & Special Collections, Goldfarb Library, Brandeis University, The Jospeh Heller Estate.

CHAPTER 7 LISTS © The Saul Steinberg Foundation/Artists Rights Society (ARS), New York. Cover reprinted with permission of The New Yorker magazine. All rights reserved. *Method of Medicine*, Al-Zahrawi, image National Library of the Kingdom of Morocco. Excerpt from *Selected Non-Fictions* by Jorge Luis Borges, edited by Eliot Weinberger, copyright © 1999 by Maria Kodama; translation copyright © 1999 by Penguin Random House LLC. Used by permission of Viking Books, an imprint of Penguin Publishing Group, a division of Penguin Random House LLC. All rights reserved. Aelius Donatus, *Ars Maior*, image Paris Bibliothèque nationale de France. Postcard © 2020 Estate of Ad Reinhardt and Artists Rights Society (ARS), New York. "How Does a Cat Fall on its Feet?" © Donald McDonald, *The New Scientist*, 1960. *Discovering Bamboo*, Jackie Heinricher © 2006 BooShoot Gardens Verb List © 2020 Richard Serra/Artists Rights Society (ARS), New York. Digital Image © The Museum of ModernArt/Licensed by SCALA/Art Resource, NY *The Art of Fiction* © David Lodge, 1992. All Rights Reserved

CHAPTER 8 REMODELING PRESENTATIONS, MEETINGS, TEACHING, BOOK PUBLISHING "Waymo can't use its self-driving forecasts, but it can use Uber's, judge rules," Cyrus Farivar, *ArsTechnica* © Conde Nast © 2012 Jeff Bezos

PUBLISHED BY GRAPHICS PRESS LLC
POST OFFICE BOX 430
CHESHIRE, CONNECTICUT, USA 06410
WWW.EDWARDTUFTE.COM